Processing Random Data
Statistics for Engineers and Scientists

Processing Random Data

Statistics for Engineers and Scientists

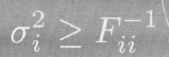

$$F_{ij} = \left\langle -\frac{\partial^2 \log p(\{d_n\}|\boldsymbol{\alpha})}{\partial \alpha_i \partial \alpha_i} \right\rangle$$

$$\sigma_i^2 \geq F_{ii}^{-1}.$$

Robert V. Edwards

Chemical Engineering Department
Case Western Reserve University, USA

World Scientific

NEW JERSEY · LONDON · SINGAPORE · BEIJING · SHANGHAI · HONG KONG · TAIPEI · CHENNAI

Published by

World Scientific Publishing Co. Pte. Ltd.

5 Toh Tuck Link, Singapore 596224

USA office: 27 Warren Street, Suite 401-402, Hackensack, NJ 07601

UK office: 57 Shelton Street, Covent Garden, London WC2H 9HE

British Library Cataloguing-in-Publication Data
A catalogue record for this book is available from the British Library.

PROCESSING RANDOM DATA
Statistics for Engineers and Scientists

ISBN-13 978-981-256-834-2
ISBN-10 981-256-834-4

Editor: Tjan Kwang Wei

Printed in Singapore

Dedication

This book is dedicated to the dozens of students and colleagues who have worked with me to figure out the concepts contained here. I would like to particularly thank Lars Lading who forced me to consider rigorously what it meant to make a measurement when there was noise present. Finally, I would like to thank my wife Anne, who kept me on the task of writing and editing the book.

Preface

These notes are being written to address a need that I perceive for an introductory statistics text that is geared toward engineers rather than epidemiologists. The difference in the need is based on the fact that engineers are often dealing with data from experiments where there is a mathematical model describing the relation between the input to the experiment and the output. For example, one might try to estimate a heat transfer coefficient by examining the temperature distribution along a rod heated at one end. There is a model connecting this temperature distribution to the heat transfer coefficient. The experimentalist will want to compare the theoretical distribution to the measured one to estimate the heat transfer coefficient. This is in contrast to a typical experiment in the social sciences where the only clear relation known is the concept of cause and effect. This text will start out with the classical non-parametric statistics concepts. The concepts introduced in that section will be expanded on in the sections dealing with model based data.

In the first sections of the text I introduce the concepts used to do computations in statistics. Examples are probability density functions and expected values. Some of the more common probability density functions will be given and used in demonstrations. I hope I will do so with enough completeness that the reader can extend the methodology to other probability densities of interest. In the second section, I will use the concepts to estimate the results of real measurements of finite numbers of random numbers. Next, I will describe methods for dealing with time series. Following that, I describe various parameter estimation schemes that can be used to estimate parameters from typical science and engineering experiments like the heat transfer experiment mentioned above. Finally there is a section on random sampling of time series.

Contents

Chapter 1

Random Variables

1.1 Basic Concepts

In this chapter, we concern ourselves with the basic concepts involved in dealing with random numbers. The first question to be dealt with is, "What is a random number?" I will give plausible answers to this question rather than a formal definition. This chapter will deal with the formal calculus of statistics. Chapter 2 will use the concepts developed to do useful calculations.

A set of random numbers is a set of numbers for which knowledge of any subset of the numbers will not tell you *with certainty* the value of the next one sampled from the set. For certain sets of numbers like the outcome of the roll of a die, the possibilities are limited and you may occasionally actually guess the result of the next throw. The point is that you cannot do it with certainty. Otherwise, playing dice would not be gambling!

A typical situation in science or engineering is that you get a set of measurements from a system. The next measurement is not certain, no matter how many measurements you took before and no matter how accurately you took those numbers. This situation is typified by a process where you put a set of small pieces of paper in a box, each containing a different number. If the contents of the box are mixed up, you should not be able to predict the next number drawn by looking at the previous number drawn. If you know that no number is repeated, you are in the position of exactly knowing what numbers will *not* be drawn, but you cannot be certain what the next number will be.

In engineering and science, you are often in the position of having a system where the inputs are constant, but the output(s) fluctuate in a random manner. For instance, you attempt to measure the signal strength from a cell phone 5 km away. The measured signal strength will vary with time because of the events occurring in the atmosphere between you and the cell phone. You will not be able to predict exactly how those fluctuations will occur. The field strength measurements are interfered with by the fluctuations. On the other hand, if you hook up a flow system to send water at a constant rate though a pipe, you will find that under certain conditions, everything you can control is constant, but the velocity of the flow at a given point is fluctuating. This flow condition is called turbulence and is often the subject of study itself.

The examples in the previous two paragraphs are typical in that the process that connects the inputs to the outputs is where the randomness is generated.

I have seen definitions that describe a set of random numbers as a set of numbers that takes at least as many bits to describe as is contained in the set. The easiest way to understand this concept is to imagine that the set of random numbers is all decoded into bits (*viz.* 6 = 0 1 1 0) and then strung all together to make one big binary word. If it were possible to find repeating sets of bits in the binary word, then we could describe the set of numbers with fewer bits than appear in the original set. Let's make this more concrete by a simple example. Consider a set of coin flips of an "honest" coin. We will call "heads" 1, and "tails" 0. Make a 100 bit binary word by flipping the coin 100 times and recording the result of the nth trial in the nth place in a 100 bit binary word. If the coin were truly "honest", you would need all 100 bits to be guaranteed that you could uniquely describe the results of all your trials. Modern data compression schemes such as used in MP3 depend on the accidental redundancy in any particular realization of a set of 100 coin flips, but this does not change the need for all 100 bits to be guaranteed a unique representation.

In many instances, random numbers selected sequentially are not related to each other in the sense that if one particular number is larger than the mean, the next number could be higher or lower than the mean. In such a situation, it doesn't matter in what order you deal with the numbers. For instance, the errors in the adjacent measurements of the temperature on the rod mentioned above are usually not related to each other.

On the other hand, there are situations where the random numbers are related to each other. If a number is larger than the mean, the numbers next to it are likely to also be larger than the mean. Here, it matters what order the numbers are sampled in. This leads us to the concept of *the time series*. A time series is a set of random numbers where the order matters.

It is important to know which one is first, second, etc. This condition can arise when measuring the fluctuations in the field strength of a cell phone, if the time scale is small enough. Events in the atmosphere that cause the fluctuations can only change but so fast.

The result of a calculation involving random numbers is called a *statistic*. For instance, if you add together four samples of a set of random numbers, the resulting statistic would also be a random number. The study of the results of computations using random numbers is called *statistics*. A statistic well known to baseball fans in the US is the batting average. It is computed by dividing the number of hits that result in the player getting on base by the number of times the player attempts to bat times 1000. It is taken as a measure of the player's ability. Since no player has ever hit safely every time at bat through a season, the chances of the player hitting safely is assumed be a random number proportional to his or her ability.

In studying statistics, you will have to define some strange, idealized objects that are never seen in real life. For instance, there is an implicit assumption made when doing the calculus of statistics that the numbers you are dealing with come from a common source called a *population*. In principle, there are an infinite number of elements in the population, but you can only deal with a sample of them. For instance, if you measure the diameter of axles made by some machine, the set of measurements made up from sample diameters of the machine output would be a sample of the population of all possible axles this particular machine could make. Implicit in this assumption is that the machine could have made an infinite number of samples without wear and tear in the cutting tools etc. You are dealing with the infinite variety of axles the machine could make in its current configuration. The settings of the tool define the state that creates the population.

If you have two machines making axles and you keep the measurements of the diameters separate, then you would be dealing with two populations. If, on the other hand, you just throw axles from both machines in a pile and make measurements later, without noting which machine the sample came from, you are back to dealing with a single population.

Suppose you are put in charge of making sure that all the axles shipped to a customer are within a certain diameter range. You would then measure every axle made by the machine. Suppose, also, that you get one axle that is outside the allowed range when all the previous ones had been within range. How do you tell if it is "just a fluke" or if you have evidence that the machine's condition has changed and it is behaving in a substandard manner? Or, can you tell the difference between the two alternatives presented above?

Further, suppose you are put in charge of measuring the heat transfer coefficient for a new design for a compact heat sink designed by your company. If you measure the heat transfer coefficient for the same sample more than once, I guarantee that each measured heat transfer coefficient will be different. How close do you think you got with any of the measurements to the "right" value? How much variation in the results between experiments is reasonable?

Answers to the above and other related problems are derived through a formal calculus of statistics. This calculus involves manipulation of the formal objects hinted at above such as the probability distribution functions, the expected value, etc.

We start by examining the idealization of the average of a sample of random numbers. In statistics texts, what we informally call the average of something, is an approximation to the formal concept of the *expected value*.

1.2   Expected Values

If x is the outcome of some experiment with a random component, the practical definition of the expected value, $\langle x \rangle$, is as the average of an *infinite* number of trials of the experiment. This type of averaging is performed by doing the same experiment many times. It is called *ensemble averaging*. A more mathematical definition will be given below.

1.3   Probability Distribution Functions

Suppose an experiment could only have a finite number of outcomes. An example is the rolling of unloaded dice. Let the nth outcome of the rolling of a die be denoted by x_n. The average outcome can be computed by adding up the values of all the events and then dividing by the total number of events. Alternatively, since there are only a finite number of possibilities for the outcome, the same calculation could be performed by measuring the number of times each value of x_n occurs. Let $P(x_n)$ be the number of times a particular value of x_n occurred. A typical result from rolling a 6 sided die 20 times is the sequence

$$(3, 1, 4, 6, 6, 6, 1, 3, 6, 1, 2, 1, 1, 1, 2, 1, 2, 3, 4, 3)$$

Here, $P(1) = 7$, $P(2) = 3$, $P(3) = 4$, $P(4) = 2$, $P(5) = 0$, $P(6) = 4$.

The average of this sequence can be computed directly by adding up the numbers in the sequence and dividing by 20. The result is 2.85. You can also compute the average by grouping like results as we did in defining $P()$. The sum of the results can be computed by adding up the products $x_n P(x_n)$. The number of results is $\sum_n P(x_n)$. The average can then be computed as

$$\bar{x} = \frac{\sum_n x_n P(x_n)}{\sum_n P(x_n)} \, . \tag{1.3.1}$$

In the example sequence above, the average is computed

$$\bar{x} = \frac{1 \times 7 + 2 \times 3 + 3 \times 4 + 4 \times 2 + 5 \times 0 + 6 \times 4}{7 + 3 + 4 + 2 + 0 + 4} = \frac{57}{20} = 2.85 \, .$$

For a large number of rolls of a single die, a typical set of measurements would give a $P(\ )$ like that shown in Figure 1.1.

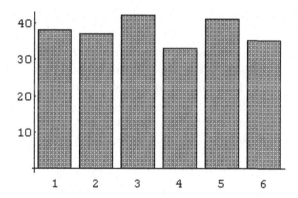

Fig. 1.1   Results of experiment rolling one die.

Experience has shown that if the experiment (rolling the die) were repeated a very large number of times, the ratio $P(x_n) / \sum P(x'_n)$ would tend to $1/6$. This ratio is called the *probability of the result* x_n. It should be clear that the probability is bounded by 0 and 1. If the result never occurs (probability is 0), we don't care about it. If it occurs all the time, (probability is 1), the process is not random.

Hence, in this case, $\langle x \rangle$, the infinite average, can be computed by using the limit of the weighting, assuming an infinite number of samples.

$$\langle x \rangle = \lim_{N \to \infty} \frac{\sum P(x)x}{\sum P(x')} = \frac{1}{6} \sum_{x=1}^{6} x = \frac{21}{6} = 3.5 \,. \qquad (1.3.2)$$

The *expected* value is 3.5; however, note that the value is not the most likely value! In fact, you can't roll a die and get a value of 3.5. In this case, the expected value should be viewed as being the limiting value of the average of the outcomes, not as a statement about what outcomes are expected from each trial. For the example given here, the expected value is not a particularly useful concept — at least it gives incomplete knowledge about the data set. Please hang in there. More useful concepts will be given shortly.

The limit of the function

$$P(x_n) \Big/ \sum_{k} P(x_k)$$

as the number of samples goes to infinity is called the *probability distribution function* (pdf) for the event x_n. This function is denoted $p(x_n)$. It can be seen to be the expected relative frequency of event x_n in any sampling of the output. In this case, it can be computed *a priori* by assuming that the probability of an outcome is equal to the number of ways the outcome can occur, divided by the total number of possible outcomes. For this example, each outcome could only occur in one way and there were 6 equally possible outcomes. The *a priori* probability of any outcome of rolling a die is thus 1/6. If the pdf is determined by measurement, it is called the *a posteriori* pdf.

The mathematical definition of the expected value for a random variable with a discrete output should now be apparent. It is defined in terms of the probability density function, *viz.*

$$\langle x \rangle = \sum_{\text{all possible } x} x p(x) \,.$$

I will deal with random variable with a continuous output below.

If the function $p(x)$ had a peak in it, the value of $p(x)$ at the peak would be *the most likely outcome*.

Consider a more complicated experiment, rolling two dice at the same time and adding the result. Here the possible sums range from 2 to 12.

There are $6 \times 6 = 36$ possible outcomes.

There is only one way to get 2 or 12, $1 + 1$ or $6 + 6$, so the *a priori* probability of each of these is $1/36$. There are more ways to get 3, $1 + 2$, $2 + 1$. The *a priori* probability of 3 and 11 is $1/18$. Continuing this argument, the table of *a priori* probabilities shown below would be constructed.

Table 1

Table of probabilities for rolling two dice

x	$p(x)$	x	$p(x)$
2	1/36	8	5/36
3	1/18	9	1/9
4	1/12	10	1/12
5	1/9	11	1/18
6	5/36	12	1/36
7	1/6		

This $p(x)$ peaks at 7. Monopoly^{TM} players and gamblers take note! The expected value of the experiment is given by

$$\langle x \rangle = \sum p(x)x = 2/36 + 3/18 + 4/12 + 5/9 + 30/36 + 7/6$$
$$+ 40/36 + 9/9 + 10/12 + 11/18 + 12/36 = 7.00.$$

Here, the most likely outcome and the expected outcome are the same. If you were to bet money on an outcome of the roll of dice, you could expect to lose money. If, for instance, you bet on 7, the chances are $1/6$ that it will occur. The chance that it won't is $5/6$.

Exercise: Try constructing the equivalent to the table above, if you have two 5 sided dice (There are such things!).

Exercise: If you have access to the spreadsheet program Excel®, you can examine the behavior of a pair of dice without wearing out your desktop. Excel® has a feature under the "Options" box in the Toolbar called "Analysis Tools." If you now double click Analysis Tools, the first time there will be a pause as the add-on program is loaded. When the menu box comes up you will see a list of options, containing among other things, "Histogram" and "Random Number Generation." Double click Random Number Generation.

Notice that one box is labeled "Distribution." If you double click the arrow on the right of this box, a list will appear that contains among other things, "Normal" and "Discrete." Click Discrete. Now, click "Number of

Variables." Here, type in 2. Now click "Number of Random Numbers" and insert the number of rolls you want. I suggest 100 for the first time. The prescription here should generate a 2 column by 100 row output. Finally, you need to tell the random number generator what range of numbers and their probabilities to use. You need to have made a column of the possible outcomes of the roll of a die, i.e. 1,2 3,... to 6. In a parallel column, type 1/6. 1/6,... For the last entry, type 1 — Sum of the other probabilities. My version of Excel$^{®}$ has a little glitch. It insists that the sum of the probabilities be 1, but its own arithmetic is not accurate enough to make the sum of the 6, 1/6s close enough to 1 for it. Click "Parameters" and insert the range where you have already inserted the properties, e.g. E1:F6. Now, click onto the "Output Range" box. In here you should insert the starting address of the column you wish to contain the random number output, e.g. A1. If you now click "OK," Excel$^{®}$ will fill the indicated range with random integers, from 1 to 6, all with equal probability. You can now compute the sum of each pair of random numbers in the two output columns. This sum column simulates the sum of two dice.

You are now in a position to examine what the generator has wrought. First, estimate the expected value by using the Excel$^{®}$ function Average() on the original, un-summed output columns. Is the result 3.5000? Why not?

Next, find another short column in the spread sheet and put in the numbers $2, 3, 4, 5, 6, 7, 8, 9, 10, 11, 12$. I assume here that you used the "D" column, starting with 1. Now, go back into Options and click Analysis Tools. Select "Histogram." Click "Input Range." Insert the range containing the numbers you want to analyze, here C1:C100. Click "Bin Range." Here you insert the histogram points, D1:D11. When you click "Output," insert the range you want the output to appear in, say G1. Before you do anything else, click off the box "Pareto" and the box "Cumulative." If those boxes are cleared, click "OK." If your machine is set up like mine, you will see a chart flash by and then disappear. You should see the list of the number of random numbers in each range listed in a column, starting with G1. Depending on the version of Excel$^{®}$ you are using, the chart will either appear next to the histogram or on a separate sheet. If it is on a separate sheet, and you wish to see the graph, go up and click "Window." Under this should be listed something called "Chart1." Click this and you should then see your histogram.

Compare your measured result to Table 1. According to Table 1, we should expect 100/36 points in the "2" bin and 100/18 in the "3" bin. Typically, the results should not be exactly what was predicted. There should be more numbers in the "6" and "7" boxes than in the 2 box, but the result should not be exactly the prediction of Table 1. As with anything involving random numbers, there are variations in the output from what is expected. If Mother Nature wanted you to know things precisely, she never would have put them in the form of random numbers. But, as we shall see in this text, there are patterns and structures that we can find.

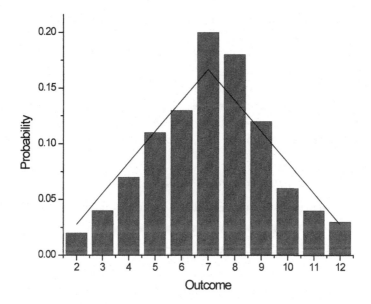

Fig. 1.2 Typical result of relative frequency test. The bars are the measurements and the line is the theory.

Now that you are sophisticated in Excel, create 2 sets of 1000 random integers and look again at the histogram for the sum. This should look a lot more like Table 1. In general, in measurement of random variables, more is better. More on this later. Save this spreadsheet and remember how you made it.

I should point out here that random numbers come in two flavors: discrete and continuous. An example of a system with a discrete outcome is the result of rolling a die. An example of a continuous random outcome would be the voltage sampled from a noisy electrical circuit. Typically, a noise voltage could be any number between -10v and 10v.

For a random variable with a continuous output, the probability for finding an output in the interval x, dx is defined as

$$p(x)dx\,. \tag{1.3.4}$$

The definition of the expected value for a continuous random variable that ranges from a to b is

$$\langle x \rangle = \int_a^b p(x)x\,dx\,. \tag{1.3.5}$$

Distinct random numbers also have a further division into finite outcomes and infinite possible outcomes. The number of counts measured in a set time for a radioactive decay experiment must be an integer, but in principle, could be any number between zero and infinity.

1.3.1   *Properties of probability density functions*

Probability distribution functions have some interesting and important properties:

$$\sum_{\text{all } x} p(x) = 1\,, \tag{1.3.6}$$

for discrete variables and

$$\int_{\text{all } x} p(x)dx = 1\,, \tag{1.3.7}$$

for continuous variables. In other words, the sum of the probabilities of all the possibilities is one.

For a discrete random variable, if $p(x)$ is the probability of an event happening, $1 - p(x)$ is the probability that it won't happen. For the continuous case,

$$\int_a^b p(x)dx \tag{1.3.8}$$

is the probability of an outcome between a and b,

$$1 - \int_a^b p(x)dx \qquad (1.3.9)$$

is the probability that the outcome will not be between a and b.

If $p(x_1)$ and $p(x_2)$ are the respective probabilities of two *independent* events x_1 and x_2, the probability of x_1 *and* x_2 is

$$p(x_1)p(x_2)\,.$$

If the probability of x_1 *and* x_2 is $p(x_1)p(x_2)$, the variables x_1 and x_2 are said to be *statistically independent* variables.

If

$$p_1(x_1) \text{ and } p_2(x_2) \qquad (1.3.10)$$

are the respective probabilities of two mutually exclusive events x_1 and x_2, the probability of x_1 *or* x_2 is

$$p_1(x_1) + p_2(x_2)\,. \qquad (1.3.11)$$

Note that 1.3.9 could have been derived from this property.

If $p_x(x)$ is the pdf for random variable x and $p_y(y)$ is the pdf for the independent random variable y, the pdf for the sum $z = x + y$ is given by

$$p(z) = \sum_x p_y(z - x)p_x(x) \qquad (1.3.12)$$

for discrete variables and,

$$p(z) = \int_x p_y(z - x)p_x(x)dx \qquad (1.3.13)$$

for continuous variables.

This property was used implicitly in the derivation of Table 1.

1.4   The Expected Value Process

You may compute the expected value of any function of x, $f(x)$ as follows,

$$\langle f(x) \rangle = \sum_x f(x)p(x)\,, \qquad (1.4.1)$$

for discrete variables.

In this case of a continuous variable, the expected value is computed by

$$\langle f(x) \rangle = \int_{\text{all } x} f(x)p(x) \, dx \,, \tag{1.4.2}$$

$$\langle x \rangle = \int_{\text{all } x} xp(x) \, dx \,. \tag{1.4.3}$$

An example of the use of such a distribution is shown below:

Let V be a fluctuating, random voltage applied to a circuit. The instantaneous power is V^2/R, where R is the resistance of the circuit. The expected average power is given by

$$\int \frac{p(V)V^2}{R} dV \,.$$

The process of computing the expected value of something is a linear operation; *i.e.*,

$$\langle \alpha f(x) + \beta g(x) \rangle = \alpha \langle f(x) \rangle + \beta \langle g(x) \rangle \,, \tag{1.4.4}$$

where α and β are fixed numbers and $f(x)$ and $g(x)$ are any functions of the random variable x. The proof is rather straightforward. Take the continuous case.

$$\begin{aligned}
\langle \alpha f(x) + \beta g(x) \rangle &= \int_{\text{all } x} p(x)(\alpha f(x) + \beta g(x)) dx \\
&= \int_{\text{all } x} \alpha p(x)f(x) dx + \int_{\text{all } x} \beta p(x)g(x) dx \\
&= \alpha \int_{\text{all } x} p(x)f(x) dx + \beta \int_{\text{all } x} p(x)g(x) dx \\
&= \alpha \langle f(x) \rangle + \beta \langle g(x) \rangle \,.
\end{aligned}$$

This property will be used often in later sections.

1.5 Variance and Standard Deviation

You often want to measure the spread of the likely outcomes about the mean. How wild a random variable is it? There are many ways to calculate a measure of the spread of a random variable, but it is usually done by computing the "variance", σ^2, of the probability distribution function.

$$\langle (x - \langle x \rangle)^2 \rangle \equiv \sigma^2 = \sum (x - \langle x \rangle)^2 p(x) \,. \tag{1.5.1}$$

If we expand the formula for calculating the expected value of a function of x shown above, we can express the calculation of the variance in another way.

$$\sigma^2 = \sum (x^2 - 2\langle x \rangle x + \langle x \rangle^2)p(x)$$

$$\sigma^2 = \sum x^2 p(x) - 2\langle x \rangle \sum x p(x) + \langle x \rangle^2 \sum p(x)$$

$$\sigma^2 = \langle x^2 \rangle - 2\langle x \rangle \langle x \rangle + \langle x \rangle^2$$

$$\sigma^2 = \langle x^2 \rangle - \langle x \rangle^2 . \tag{1.5.2}$$

The quantity $\langle x^2 \rangle$ is the expected value of x^2. For the pair of dice,

$$\langle x^2 \rangle = 54.833, \ \langle x \rangle^2 = 49$$

$$\sigma_x^2 = 5.833$$

$$\sigma_x = 2.415 .$$

There are other measures for the deviation of a random variable from its expected value such as $\langle |x - \langle x \rangle| \rangle$, but the variance is usually the easiest to compute.

The square root of the variance, the *standard deviation*, of the probability function is the desired measure of the width of the function. This measure is particularly useful if the payoff from the outcome is a function of how much a given outcome differs from the expected value. Typically, if $\sigma/t\langle x \rangle$ is small, the distribution is said to be narrow. The variations about the expected value are not large compared to the expected.

These concepts can be extended to any experiment where the output has a random component.

Likewise, for a continuous variable,

$$\sigma^2 = \int (x - \langle x \rangle)^2 p(x) dx = \langle x^2 \rangle - \langle x \rangle^2 . \tag{1.5.3}$$

1.6   Moments and Moment Generating Functions

The nth moment of a random variable is defined as $\langle x^n \rangle$. The nth central moment is defined as

$$\langle (x - \langle x \rangle)^n \rangle \,.$$

As we have seen, knowledge of the first moment and the second central moment (the variance) gives us some information about the shape of the pdf, but the description so far is incomplete. For instance, the mean and the variance give no clue as to whether the pdf is symmetrical about some point.

It is easy to show that, if the pdf is symmetrical about the expected value, the odd central moments are zero. Hence a measure of the asymmetry of a distribution can be given by the normalized third central moment, *viz.*,

$$S = \langle (x - \langle x \rangle)^3 \rangle / \sigma^3 \,. \tag{1.6.1}$$

If this quantity is non-zero, the distribution is not symmetric. A positive value indicates a distribution with a longer positive tail than the negative tail. A negative value indicates the reverse.

A complete knowledge of the moments is equivalent to a complete knowledge of the pdf and vice versa. One way this can be seen is by considering the *Moment Generating Function*, or the *Characteristic Function* of the pdf, defined as follows:

$$Q_x(s) \equiv \int e^{-sx} p(x) dx = \langle e^{-sx} \rangle \,. \tag{1.6.2}$$

If the range of x is contained within $(0, \infty)$, s is usually taken as real and the integration is the Laplace transform of the pdf. In this case, Q_x is called the moment generating function.

Expand e^{-sx} in a Taylor's series and integrate,

$$Q_x(s) = \int_0^\infty (1 - sx + \frac{s^2 x^2}{2} - \frac{s^3 x^3}{3 \times 2} + \dots) p(x) dx$$

$$= \int 1 p(x) dx - s \int x p(x) dx + \frac{s^2}{2} \int x^2 p(x) dx - \frac{s^3}{3 \times 2} \int x^3 p(x) dx \dots$$

$$= 1 - s \langle x \rangle + \frac{s^2 \langle x^2 \rangle}{2} - \frac{s^3 \langle x^3 \rangle}{3 \times 2} + \cdots \,.$$

The nth moment is given by

$$\langle x^n \rangle = (-1)^n \frac{d^n Q_x(s)}{ds^n}\Bigg|_{s=0}. \tag{1.6.3}$$

If the range of x includes negative numbers, it is customary to keep s real, but to substitute is for s, $i = \sqrt{-1}$, in 1.6.2, and extend the integration from $(-\infty, \infty)$. In this case, you are doing a Fourier Transform and the resulting function is called the characteristic function. The formula for the moments is essentially the same, except that (i^n) is substituted for $(-1)^n$.

Similar calculations can be done for the central moments.

1.7 Common Types of Distributions

In this section I will describe some of the pdfs that often appear in the literature. They all have names. The pdf indicated above for the sum of the result of rolling three dice does not have a name, but is a well defined distribution. So please understand that the distributions described below are only a representative sample of the class of distributions.

1.7.1 *Uniform distribution*

This distribution is the simplest of all. It can describe both continuous and discrete outputs. In either case, the pdf is either a constant or zero. For the discrete case,

$$p(x_i) = \begin{cases} 1/N; & 1 \le i \le N \\ 0; & \text{otherwise} \end{cases}. \tag{1.7.1}$$

For the continuous case,

$$p(x) = \begin{cases} 1/(c - b); & b \le x \le c \\ 0; & \text{otherwise} \end{cases}. \tag{1.7.2}$$

Using either the moment generating function for the continuous uniform distribution or using the definition of the moments directly computed from the pdf, you would get

$$\langle x \rangle = \frac{c + b}{2} \quad \text{and} \quad \langle x^2 \rangle = \frac{b^2 + bc + c^2}{3}.$$

The variance is thus

$$\sigma^2 = \langle x^2 \rangle - \langle x \rangle^2 = \frac{(c-b)^2}{12}.$$

There are uniform generators in Excel® and Mathematica® that seem to work well. Understand that random number generators on computers are all so-called pseudo-random generators. For example, if you start the generator out at the same value (called the random seed), the generator will always give the same output sequence. A good pseudo-random generator will not display an obvious correlation between the output values and will closely mimic the pdf of the distribution. Typically, a computer random generator will give you the choice of picking the random seed or it will generate its own, sometimes by manipulating a value obtained from the computer's clock.

1.7.2 *Binomial distribution*

Consider a situation where you have n attempts to shoot a foul shot in basketball. Let μ be the probability for each shot that you make the shot. The probability that you will miss the shot is $1 - \mu$. We also assume that the probability for each shot is the same as all the others and independent. It should be immediately obvious that the chances that you will make none of the n shots is given by $p(0) = (1 - \mu)^n$ (I know you think this is impossible, but I include it for mathematical completeness.)

The probability, $p(1)$, of making exactly one shot out of the total is given by $p(1)$ = (the probability of making the first shot) × (probability of making no other shot) + (the probability of making the second shot) × (the probability of making no other shot) + ...

$$p(1) = \underbrace{\mu(1-\mu)^{n-1} + \mu(1-\mu)^{n-1} + \ldots}_{(n \text{ times})}$$

$$p(1) = n\mu(1-\mu)^{n-1}.$$

By a similar argument, you can show that $p(2)$ = (number of ways of making two shots in n tries without regard to order) times (The number of pairs in n) × $\mu^2(1-\mu)^{n-2}$.

The symbol

$$\binom{n}{m} = \frac{n!}{(n-m)!m!}$$

denotes the number of ways of putting m things in n places without regard to order. It is called the Binomial Coefficient.

$$p(2) = \binom{n}{2} \mu^2 (1-\mu)^{n-2} . \tag{1.7.3}$$

In general,

$$p(m) = \binom{n}{m} \mu^m (1-\mu)^{n-m} . \tag{1.7.4}$$

This is the probability of making m shots in n tries if the probability for each shot is μ. It is called the *binomial distribution*.

The definition of the binomial distribution may seem a bit arcane and not really related to everyday life, but, in point of fact, the binomial distribution is by far the most often used distribution that affects your life. It is the distribution used to tell you how dangerous a pesticide might be. It is used to tell you how many people would vote for a given candidate. It is used to determine which TV shows are the most popular. A direct consequence of this is that the binomial distribution is often the most abused distribution. It is used to tell you that coffee is bad for you. Then, it is used in another study to tell you that coffee is good for you... Usually, the problem with clearly interpreting survey results is knowing whether the sample taken was really representative of the entire population you are extrapolating to. For instance, a telephone poll by definition does not sample people without telephones. It is conceivable that in some situations people without telephones will have a different response to a survey question than someone with a telephone. The reader may skip directly to the definitions of other distributions if you are not interested in the details of how pdfs like the binomial distribution help inform us about what happens in the real world.

A situation in which the binomial distribution is often used is when you know there are n events in an interval L. You wish to know the probability for m events in a sub-interval L_s. If the probability of an event is spread equally over the interval, $\mu = L_s/L$. The probability of m events in the interval is given by 1.7.4. The parameters n, m, and μ are used in exactly the same way.

Suppose you select 11 numbers from a continuous uniform random number population on the interval $[0,1]$. The probability of a number being in the interval, $[0,0.25]$ is $1/4$. What is the probability that only 2 of the 11 random numbers ended up in the interval $[0,0.25]$?

$$p(2) = \binom{11}{2} (1/4)^2 (3/4)^9 = 0.26.$$

The binomial distribution is most often encountered when you have a known number of independent trials of a phenomenon whose output can be one of two choices. For instance, if you do n trials where the outcome can be 1 or 2 and the probability of each 1 is μ, the probability of encountering m type one events in the trial is given by 1.7.4.

The binomial distribution is also appropriate when you have an experiment where n outcomes are distributed among k possibilities. If the probability for the pth event is μ_p, the probability of finding m_p events of the type p is also given by 1.7.4, *viz.*,

$$p(m_p) = \binom{n}{m_p} \mu_p^{m_p} (1 - \mu_p)^{n - m_p}. \tag{1.7.5}$$

This latter case is the use to which the binomial distribution will be most often used in this text. It explains the statistics for a measured pdf, a histogram. Since this is a text, there will be many examples given of measured pdfs and the results will never be exactly what is predicted to be the expected distribution. The statistics of the measured pdfs will be elucidated using the binomial distribution.

Exercise:

$$1 = (\mu + 1 - \mu)^N = \sum_{M=0}^{N} \binom{N}{M} \mu^M (1 - \mu)^{N-M}$$

Using the above definition, show that

$$\langle m \rangle = N\mu. \tag{1.7.6}$$

Hint:

$$\binom{N}{0} = 1.$$

The variance of a binomial distribution is

$$N\mu(1 - \mu) \,. \tag{1.7.7}$$

You are doing measurements of a system with ten possible outcomes. The probability of the third outcome is 0.15. If you do 200 trials, what is the probability of getting 32 occurrences of the third outcome?

1.7.2.1 *Using the binomial distribution*

A popular use of the binomial distribution occurs when you are not so much looking at how many successes you have, but are trying to figure out from knowing the number of successes what the probability of the outcome is. Suppose you figure out how to select a set of N "typical" television viewers. If you find that a certain fraction m watch "Gilligan's Island" reruns every week, how good an estimate is m/N of the probability for the entire population of television watchers? This scenario should be known to you by now. The company doing the measurement claims that they sampled about 1000 people and claim to be able to tell us the result for several hundred million television viewers. In a similar vein, doctors do epidemiological studies and tell us on alternate days that power lines cause brain cancer and then that they don't. Again, how much should you believe them?

Figuring out the quality of survey results can be done by examining the amount of variation expected for the result compared to the result itself. If we get a probability of an outcome of 0.13 and the expected variation in the measurement, the standard deviation, is 0.27, we shouldn't have a lot of confidence in the result. Using the results of the homework problems above, we can define a characteristic relative error as the expected standard deviation divided by the expected mean. By characteristic, I mean that it can generate numbers that tell you the order of magnitude of what you are dealing with. So, if the characteristic ratio of those quantities is 1%, then a variation of 10% would be considered very unusual. Likewise, if you calculate a characteristic ratio of 10%, then a 1% variation would be likewise rare.

According to the homework

$$\frac{\text{standard deviation}}{\text{mean}} = \frac{\sqrt{N\mu(1 - \mu)}}{N\mu} = \sqrt{\frac{(1 - \mu)}{N\mu}} \,.$$

If you are doing election polling, it is only worth it if $\mu \approx 0.5$, i.e. both candidates have a possibility to win. In that case the relative error is approximately $\sqrt{1/N}$. So, if you go out and get results from 1000 "typical" voters, the relative error should be about 3%. If you poll 10,000 voters instead, your relative error only goes down to 1%. Typically, national polls use on the order of 1000 samples. It tends not to be worth it to go the extra mile and use 10,000 or more samples.

If you are trying to estimate the percentage that are watching "Gilligan's Island" daily, that should be less than 5%, so the relative error for a sample of 1000 people will be

$$\sqrt{0.95/0.05 * 1000} = \sqrt{.019} = 0.14 \,.$$

In this case you get a 14% error with 1000 samples. You expect to get 50 positive measurements, but the laws of chance say that you could easily measure as low as 43 measurements or a high as 57 measurements. You would still come up with the result that Gilligan's Island is not very popular, but the result would not be as relatively accurate as for an event whose probability was near 0.5.

I will do one more example before moving on to numerical experimentation. Assume you are trying to discover whether a local plant's effluent is causing some rare disease. In a situation like this, you know the background level of disease occurrence. This would be the random occurrence of the disease without a specific cause. By the definition of "rare" condition, the probability is small. For sake of argument, let us assume that $\mu = 0.001$. If you found 2 occurrences of the disease in 500 samples, do you have good enough evidence to go to court and stop the plant from operating?

Homework: Calculate the probability of measuring a 0.4% rate if the actual rate was 0.1%. Is the probability low enough to make it improbable that the conclusion that the plant was causing a problem was in error?

1.7.3   *Poisson distribution*

Consider a situation where a line of length L is divided into n segments by cuts of length Δx. Let the probability of one particle being in a line segment be given by,

$$p = \mu \Delta x \,.$$

Using the binomial distribution just derived, the probability of there being exactly m particles in the interval L is given by

$$p(m) = \binom{n}{m} (\mu\Delta x)^m (1 - \mu\Delta x)^{n-m}$$

$$= \binom{n}{m} (\mu L/n)^m (1 - \mu L/n)^{n-m}. \tag{1.7.9}$$

If you let $\Delta x(= L/n)$ shrink to zero by letting n go to infinity while keeping the length constant, it is easy to show that

$$\lim_{n\to\infty} p(m) = \frac{(\mu L)^m}{m!} e^{-\mu L}. \tag{1.7.10}$$

This is the probability distribution for m events occurring in the interval, if the expected number of events is μL. It is called the *Poisson distribution*. This distribution occurs in many physical situations where the process has an output consisting of the sum of discrete, independent events. Photon counting experiments or radioactive experiments are typical situations where this distribution occurs. Recall that the binomial distribution is used when you know how many outcomes you have and you want to know how they probably are distributed. The Poisson distribution is used when you know the average number of outcomes and you want to know what the probability of a particular outcome is.

Computing the Poisson distribution as a limit of the Binomial distribution hints that in the limit of large n, we can approximate a Binomial Distribution with a Poisson distribution, *i.e.*

$$p(m) = \binom{n}{m} \mu^m (1 - \mu)^{n-m} \approx \frac{(\mu n)^m}{m!} \exp[-\mu n].$$

I often use this approximation when doing calculations on histograms. For example, calculate the probability of there being 10 counts in an interval for a uniform distribution divided into 100 bins and 1000 trials. The probability per bin is 0.01.

$$p(10) = \binom{1000}{10} (0.01)^{10}(1 - 0.01)^{990} \approx 0.1257.$$

But,

$$\frac{(10)^{10}}{10!} \exp[-10] \approx 0.1251.$$

Close enough.

Computation of the moments of this distribution is rather straight-forward.

$$\langle m \rangle = \sum_{m=0}^{\infty} \frac{m(\mu L)^m}{m!} e^{-\mu L} .$$

To simplify the notation here, let $\mu L = \lambda$.

$$= e^{-\lambda} \left(\frac{d}{d\lambda} \sum_{m=0}^{\infty} \frac{\lambda^m}{m!} \right) \lambda .$$

But the sum term is the series for e^λ, thus

$$\langle m \rangle = e^{-\lambda} \lambda \frac{d}{d\lambda} e^\lambda = \lambda$$

$$\langle m \rangle = \lambda = \mu L . \tag{1.7.11}$$

$$\langle m^2 \rangle = e^{-\lambda} \sum_{m=0}^{\infty} \frac{m^2 \lambda^m}{m!}$$

$$= e^{-\lambda} \left(\lambda^2 \sum_m \frac{m(m-1)\lambda^{m-2}}{m!} + \lambda \sum_m \frac{m\lambda^{m-1}}{m!} \right)$$

$$= e^{-\lambda} \left(\lambda^2 \frac{d^2}{d\lambda^2} e^\lambda + \lambda \frac{d}{d\lambda} e^\lambda \right)$$

$$\langle m^2 \rangle = \lambda^2 + \lambda . \tag{1.7.12}$$

Thus,

$$\langle m^2 \rangle - \langle m \rangle^2 = \lambda . \tag{1.7.13}$$

The mean and the variance of a Poisson distribution are both λ.

Example:
The arrival of electrons at an anode, or photons from a coherent source can both be described by a Poisson process. If μ is the mean rate of arrival, the expected number measured in a time Δt is $\mu \Delta t$. The probability of n electrons (or photons) arriving in time Δt is given by

$$p(n) = \frac{e^{-\mu \Delta t}(\mu \Delta t)^n}{n!} .$$

The variance of the number of arrivals in Δt seconds is $\mu \Delta t$.

A measure of the quality of the measurements S/N would be the RMS fluctuation in the measurements divided by the mean.

$$S/N = \frac{\mu \Delta t}{\sqrt{\mu \Delta t}} = \sqrt{\mu \Delta t}\,.$$

The ratio of the RMS fluctuations to the mean varies as the square root of the mean. This effect can be the primary source of noise in an experiment involving weak light or small signals. The fluctuations in the measurement rate of photons coming even from a steady source appear as a noise that can tend to obscure the signal.

Excel® and Mathematica® contain good Poisson generators.

1.7.4 *Gaussian (or normal) distribution*

This distribution is defined for continuous variables.

$$p(x)dx = \frac{1}{\sqrt{2\pi\sigma^2}} \exp\left(-\frac{(x-\mu)^2}{2\sigma^2}\right) dx\,. \tag{1.7.14}$$

The expected value of x is μ and the variance of x from the mean is σ^2, i.e.,

$$\langle (x-\mu)^2 \rangle = \sigma^2\,.$$

The other central moments of the Gaussian can easily calculated using the characteristic function.

$$Q(s) = \frac{1}{\sqrt{2\pi\sigma^2}} \int_{-\infty}^{\infty} e^{isx} e^{-\frac{x^2}{2\sigma^2}}\, dx\,.$$

$$Q(s) = e^{-s^2\sigma^2/2}\,.$$

In particular, the fourth central moment will be of use later in this text.

$$\langle \Delta x^4 \rangle = (i^4) \left.\frac{\partial^4 Q(s)}{\partial s^4}\right|_{s=0} = 3(\sigma^2)^2\,. \tag{1.7.15}$$

Likewise, it can be shown that all the odd central moments are zero.

The Gaussian distribution is very useful in light of a theorem known as the *Central Limit Theorem*. It is stated as follows: If x is the average of n independent random variables from a population of expected value μ and a standard deviation σ, in the limit as n goes to ∞, the probability density for the variable

$$z = \frac{(x - \mu)}{\sqrt{2\sigma^2/n}} \qquad (1.7.16)$$

becomes Gaussian, *i.e.*,

$$\lim n \to \infty p(z)dz = \frac{1}{\sqrt{\pi}}e^{-z^2}dz \,. \qquad (1.7.17)$$

In other words, as n gets large,

$$p(x) \to \frac{1}{\sqrt{2\pi\sigma^2/n}} \exp\left(-\frac{(x-\mu)^2}{2\sigma^2/n}\right) \,. \qquad (1.7.18)$$

There is a lot here. First, it says that if you average the samples of a random variable together, *the probability density of the resultant variable tends to a Gaussian no matter what the original probability distribution function.* Further, in the limit, the expected value of the average is the same as the expected value for the original population. Moreover, this formula claims that in the limit, the variance of the mean is less than the variance of the original population, going down by a factor of n, the number of points used in the average.

Example: Consider a system measuring the number of photons in a regular time interval such that the mean number per interval is 10. The probability distribution is Poisson so that the variance is 10. Figure 1.3 shows the pdf for a Gaussian of mean and variance of 10.0 (solid line), along with the Poisson pdf (shaded rectangles).

Recall that the distribution functions are often used as parts of sums or integrals, so that exact agreement at every point is not always needed for a useful approximation. As a rule, a Gaussian is a good approximation to a Poisson distribution if the expected value is greater than 10.

1.7.5   *Student-t distribution*

A useful distribution for statistical experiments is the Student-t distribution. It is used on the occasions where you have measured means and measured variances which are, by definition, estimates of the expected values.

$$f(t, \nu) = \frac{1}{\sqrt{\pi\nu}}\frac{\Gamma((\nu+1)/2)}{\Gamma(\nu/2)}\left(1 + \frac{t^2}{\nu}\right)^{-(\nu+1)/2} \,. \qquad (1.7.19)$$

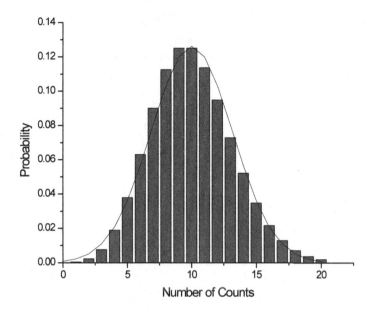

Fig. 1.3 Pdfs for a Gaussian (continuous line) and a Poisson (bars) both with a mean and variance of 10.

Here, t is the independent variable and ν is known as *the number of degrees of freedom*. Don't panic. Tables of the t-distribution are ubiquitous and there are functions in Excel® and Mathematica®.

Typically, the t is made up as follows.

$$t = \frac{\bar{x}}{s} \,,$$

where $\bar{x}$ is the measured mean obtained using N points, and s is the measured standard deviation. The number of degrees of freedom is one less than the number of points used to estimate the variance, here $N - 1$. In principle, the t distribution applies only if the original random numbers had a Gaussian distribution. However, thanks to the central limit theorem, the distribution for the averages approaches a Gaussian distribution as more and more points are used for the average. In most statistics texts, the implicit assumption is that there are enough points used that the means derived have a Gaussian distribution.

1.7.6   *Sum of two Gaussian variables*

Let z be the sum of two independent Gaussian random variables with the same mean and variance, μ and σ^2, respectively. From 1.3.13, the pdf for z is given by

$$\frac{1}{2\pi\sigma^2} \int_{-\infty}^{\infty} \exp\left(-\frac{(x-\mu)^2}{2\sigma^2}\right) \exp\left(-\frac{(z-x-\mu)^2}{2\sigma^2}\right) dx$$

$$= \frac{1}{\sqrt{4\pi\sigma^2}} \exp\left(-\frac{(z-2\mu)^2}{4\sigma^2}\right). \tag{1.7.20}$$

1.7.7   *The Chi-squared distribution*

Let z_n be a zero mean Gaussian random variable with variance one. Then define

$$\chi_\nu^2 = \sum_{n=1}^{\nu} z_n^2.$$

The pdf for this variable is

$$p(\chi_\nu^2) = \frac{1}{(2^{\frac{\nu}{2}})\Gamma(\frac{\nu}{2})}(\chi_\nu^2)^{\frac{\nu-2}{2}} \exp\left[-\frac{\chi_\nu^2}{2}\right]. \tag{1.7.21}$$

It can be shown that

$$\langle \chi_\nu^2 \rangle = \nu \tag{1.7.22}$$

and

$$\sigma_{\chi^2}^2 = 2\nu. \tag{1.7.23}$$

This distribution is most useful in dealing with the statistics of estimates of the variance. Let $z_n = (x_n - \bar{x})/\sigma$. Then,

$$\chi_{\nu-1}^2 = \sum_{n=1}^{\nu} z_n^2 = \sum_{n=1}^{\nu} \frac{(x_n - \bar{x})^2}{\sigma^2} = (\nu - 1)\frac{s^2}{\sigma^2}.$$

Note the $\nu - 1$ that comes about because you have to use the computed average to estimate the variance.

In the limit of large ν, the distribution becomes Gaussian and the chi-squared distribution can be written,

$$p(\chi_\nu^2)d\chi_\nu^2 = \frac{1}{\sqrt{4\pi\nu}}\mathrm{Exp}\left[-\frac{(\chi_\nu^2 - \nu)^2}{4\nu}\right]d\chi_\nu^2 .$$

It is straightforward to convert this to the pdf for the measured variance in the large number limit,

$$p(s^2)ds^2 = \frac{1}{\sqrt{4\pi(\nu - 1)}}\mathrm{Exp}\left[-\frac{\left(\frac{s^2}{\sigma^2}(\nu - 1) - (\nu - 1)\right)^2}{4(\nu - 1)}\right]\frac{ds^2}{\sigma^2}(\nu - 1)$$

$$p(s^2)ds^2 = \frac{1}{\sqrt{4\pi(\sigma^2)^2/(\nu - 1)}}\mathrm{Exp}\left[-\frac{\left(s^2 - \sigma^2\right)^2}{4(\sigma^2)^2/(\nu - 1)}\right]ds^2 . \qquad (1.7.24)$$

1.7.8 *The error function*

The indefinite integral of a Gaussian is a very important quantity in statistics.

$$\frac{2}{\sqrt{\pi}}\int_0^x e^{-t^2}dt = \mathrm{erf}\, x , \qquad (1.7.25)$$

the *error function*.

In terms of the distributions commonly used,

$$\frac{1}{\sqrt{2\pi\sigma^2}}\int_{-\infty}^x \exp\left(-\frac{(t - x_0')^2}{2}\frac{}{\sigma^2}\right)dt = \frac{1}{2}\left(1 + \mathrm{erf}\left[\frac{(x - x_0)}{\sqrt{2}\sigma}\right]\right) , \qquad (1.7.26)$$

where it is understood that $\mathrm{erf}(-x) = -\mathrm{erf}(x)$.

Values for the error function may be obtained form tables found in most statistics texts or from functions found in Excel® or Mathematica®.

There are many other examples of probability distribution functions, but those outlined above will be sufficient for this work. In any event, please do not adopt the standard procedure of assuming that every distribution is Gaussian!

1.8   Functions of More Than One Random Variable

Until now, we have restricted our discussion to situations involving just
one random variable. It often occurs that we have to deal with functions of
more than one random variable, where the random variables involved are
not necessarily independent. An example would be the probability of a day
in Cleveland above 90°F *and* there being a thunderstorm on that day. The
two events are linked, but they are both random variables.

The probability of 1 *and* 2 is denoted $p(1,2)$. The *conditional probability*
of 1 *if* 2 occurs is denoted $p(1|2)$. That would be the probability of a
thunderstorm *if* the temperature were above 90°F. There are several useful
theorems relating these probabilities:

1.8.1   *(Bayes theorem)*

$$p(1,2) = p(1|2)p(2) = p(2|1)p(1)\,. \tag{1.8.1}$$

The probability of 1 and 2 is the conditional probability of 1 given 2
times the probability of 2. In addition

$$p(1) = \sum_2 p(1,2)\,. \tag{1.8.2}$$

The probability distribution of the outcome $p(1)$ is obtained if you sum
$p(1,2)$ over all 2 type events. For continuous random variables, you have

$$p(x_1) = \int p(x_1, x_2)dx_2\,. \tag{1.8.3}$$

These results generalize for a set of more than two random variables,
i.e.,

$$p(x_1, \ldots, x_k) = \int p(x_1 \ldots x_n)dx_{k+1} \ldots dx_n\,. \tag{1.8.4}$$

Characteristic functions can be defined for joint pdfs:

$$Q_{xy}(s_1, s_2) = \int e^{-(s_1 x_1 + s_2 x_2)} p(x_1, x_2)dx_1 dx_2\,. \tag{1.8.5}$$

These concepts are most useful when considering functions of multiple
random variables. In general the expected value of a function of multiple

random variables is given by

$$\langle f \rangle = \iiint \cdots \int f(x_1, x_2, x_3, \ldots, x_k) p(x_1, x_2, x_3, \ldots, x_k) dx_1 dx_2 dx_3, \ldots, dx_k .$$
(1.8.6)

This formula is also useful for computing multiple moments.
Define

$$\Delta x_i \equiv x_i - \langle x_i \rangle .$$
(1.8.7)

Then

$$\langle \Delta x_1 \Delta x_2 \rangle = \iint \Delta x_1 \Delta x_2 p(x_1, x_2) dx_1 dx_2 .$$
(1.8.8)

This quantity is known as the *covariance* of x_1 and x_2. It is a measure of the correlation in the fluctuations of the two variables from their means.

An alternate method of calculating the moments is through the moment generating function of the characteristic function. The multiple moments of $p(x)$ are given by

$$\langle x_1 x_2 \ldots x_k \rangle = [-1]^k \frac{\partial^k Q(\mathbf{s})}{\partial s_1 \partial s_2 \ldots \partial s_k} .$$
(1.8.9)

Consider a simple example where $x_3 = x_1 + x_2$, where x_1 and x_2 are both random variables. We wish to know the pdf for the new random variable x_3. If we could find the pdf for x_3 and x_1, the problem could be handled rather simply. By 1.8.3,

$$p(x_3) = \int p(x_3, x_1) dx_1 .$$

It can be shown that, if x_3 is a function of x_1 and x_2,

$$p_{12}(x_1, x_2) dx_1 dx_2 = \left| \frac{\partial x_3}{\partial x_1} \right|^{-1} p_{13}(x_3, x_2) dx_3 dx_2 .$$
(1.8.10)

Here,

$$p_{12}(x_1, x_2) = p(x_3 - x_2, x_2)$$

and

$$\left| \frac{\partial x_3}{\partial x_1} \right| = 1 ,$$

so

$$p(x_3) = \int p_{12}(x_3 - x_2, x_2)dx_2 .$$

If x_1 and x_2 are independent random variables, then

$$p(x_3) = \int p_1(x_3 - x_2)p_2(x_2)dx_2 , \qquad (1.8.11)$$

the convolution of the two pdfs.

Example: Let x_1 and x_2 have a uniform distribution from 0 to 1. Then

$$p(x_3) = \int_0^{x_3} dx_2; \quad x_3 \leq 1; \quad p(x_3) = \int_{x_3-1}^1 dx_2; \quad x_3 > 1 .$$

$$p(x_3) = x_3, \quad 0 < x_3 \leq 1, \quad p(x_3) = 2 - x_3, \quad 1 < x_3 < 2; \quad p(x_3) = 0$$

otherwise.

This result is equivalent to summing over the probabilities of all combinations of x_1 and x_2 that add up to x_3.

Consider now computing the probability of $x_3 = x_1/x_2$, from the known probability of x_1 and x_2, p_{12} (x_1, x_2). We wish to make a variable change from (x_1, x_2) to (x_3, x_2). For this problem,

$$p_{12}(x_1, x_2) = p_{12}(x_3 x_2, x_2)$$

and

$$\left| \frac{\partial x_3}{\partial x_1} \right| = \left| \frac{1}{x_2} \right| .$$

Thus,

$$p(x_3) = \int p_{12}(x_3 x_2, x_2) |x_2| \, dx_2 .$$

Example: Again let x_1 and x_2 come from a uniform distribution from 0 to 1. For $x_3 \leq 1$,

$$p(x_3) = \int_0^1 x_2 dx_2 = 1/2 .$$

For $x_3 > 1$,

$$p(x_3) = \int_0^{1/x_3} x_2 dx_2 = \frac{1}{2x_3^2}.$$

Similar forms apply for the computations involving discrete event pdfs. For instance, the function in Table 1 for the probability of the sum of the results of rolling two dice can be computed from the individual pdf for each die. Taking into account that each die is independent, we get

$$p(\text{sum}) = \sum_{x_1 + x_2 = \text{sum}} p_1(\text{sum} - x_2)p_2(x_2).$$

But we know that each p_i has the same value, $1/6$. Therefore, the calculation becomes

$$p(\text{sum}) = \frac{1}{36} \times (\text{number of terms where } x_1 + x_2 = \text{sum}).$$

Try to compute the pdf for the sum of three dice.

1.8.2 *Joint Gaussian distributions*

If you have multiple zero mean random variables that have a joint Gaussian distribution, the pdf for the data set would be written

$$p(x_1, x_2, \ldots, x_n) = (2\pi)^{-n/2} |\boldsymbol{\Lambda}|^{-1/2} \exp(-0.5(\mathbf{x}) \cdot \boldsymbol{\Lambda}^{-1} \cdot \mathbf{x}), \qquad (1.8.12)$$

where $\boldsymbol{x}$ is the vector of random variables,

$$\mathbf{x} = (x_1, x_2, x_3, \ldots), \Lambda_{ij} = \langle x_i x_j \rangle,$$

and $|\boldsymbol{\Lambda}|$ is the determinant of $\boldsymbol{\Lambda}$.

The multiple moments of this distribution can be calculated through use of the characteristic function.

$$Q(s_1, s_2, \ldots, s_k) = (2\pi)_{-n/2} |\boldsymbol{\Lambda}|^{-1/2} \int_{-\infty}^{\infty} \exp(-0.5(\mathbf{x} \cdot \boldsymbol{\Lambda}^{-1} \cdot \mathbf{x}))$$

$$\times \exp(i\boldsymbol{s} \cdot \mathbf{x}) d\mathbf{x}, \qquad (1.8.13)$$

By completing the square of the exponent of the integrand, we get

$$Q(s_1, s_2, \ldots, s_k) = (2\pi)^{-n/2} |\boldsymbol{\Lambda}|^{-1/2}$$

$$\times \int_{-\infty}^{\infty} \exp(-0.5(\mathbf{x} - i\boldsymbol{\Lambda} \cdot s) \cdot \boldsymbol{\Lambda}^{-1} \cdot (\mathbf{x} - \boldsymbol{\Lambda} \cdot s))$$

$$\times \exp(-0.5 \mathbf{s} \cdot \boldsymbol{\Lambda} \cdot \mathbf{s}) d\mathbf{x} .$$

Now, performing the multiple integrations, we get

$$Q(s_1, s_2, \ldots, s_k) = \exp(-0.5 \mathbf{s} \cdot \boldsymbol{\Lambda} \cdot \mathbf{s}) . \tag{1.8.14}$$

The fourth multiple moment will later be of interest.

$$\langle x_1 x_2 x_3 x_4 \rangle = \left. \frac{\partial^4 Q}{\partial s_1 \partial s_2 \partial s_3 \partial s_4} \right|_{s_1 = s_2 = s_3 = s_4 = 0}$$

$$= \Lambda_{12} \Lambda_{34} + \Lambda_{13} \Lambda_{24} + \Lambda_{14} \Lambda_{23} .$$

Thus, the fourth central moment can be written

$$\langle x_1 x_2 x_3 x_4 \rangle = \langle x_1 x_2 \rangle \langle x_3 x_4 \rangle \langle x_1 x_3 \rangle \langle x_2 x_4 \rangle$$

$$+ \langle x_1 x_4 \rangle \langle x_2 x_3 \rangle . \tag{1.8.15}$$

If the variables are not zero mean, it can be shown that the characteristic function is given by

$$Q(s_1, s_2, \ldots, s_k) = \exp(i\mathbf{s} \cdot \boldsymbol{\mu}) \exp(-0.5 \mathbf{s} \cdot \boldsymbol{\Lambda} \cdot \mathbf{s}) , \tag{1.8.16}$$

where $\boldsymbol{\mu} = (\mu, \mu, \mu \ldots)$, $\mu = \langle x_i \rangle$.
In this case

$$\langle x_1 x_2 x_3 x_4 \rangle = \langle x_1 x_2 \rangle \langle x_3 x_4 \rangle + \langle x_1 x_3 \rangle \langle x_2 x_4 \rangle$$

$$+ \langle x_1 x_4 \rangle \langle x_2 x_3 \rangle - 2\mu^4 . \tag{1.8.17}$$

1.9   Change of Variable

This section is devoted to examining, in more detail than before, the concept of computing the probability density function of a function of a random variable x if we know the probability density function for x.

Let $y(x)$ be a monotone (either increasing or decreasing) function of x. See Figure 1.4. From this figure, it can be seen that the probability of the function shown being in the interval $y(x)$, $y(x) + dy$ is the same as the

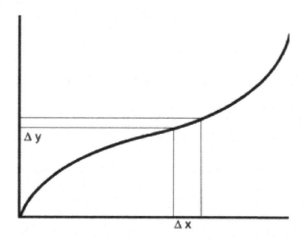

Fig. 1.4 Monotone function.

probability of x being in the interval x, $x + dx$. Further, dy is related to dx by

$$dy = \left| \frac{dy}{dx} \right| dx .$$ (1.9.1)

The absolute value sign comes about because we are only interested in the length of dy, not its sign. It follows then that the pdf for $y(x)$ is given by

$$p(y) = \frac{p_x(x(y))}{\left| \dfrac{dy}{dx} \right|} ,$$ (1.9.2)

where the pdf on the right is the pdf for x.

Example:
Suppose x is a Gaussian random variable with mean μ. What is the pdf for $\ln x$?

$$p(x) = \frac{1}{\sqrt{2\pi\sigma^2}} \exp\left(-\frac{(x - \mu)^2}{2\sigma^2} \right).$$

$$g(x) = \ln x .$$

$$\frac{dg}{dx} = \frac{1}{x} = e^{-g} \cdot x > 0 .$$

Thus,

$$p(g) = \frac{1}{\sqrt{2\pi\sigma^2}} \exp\left(g - \frac{(e^g - \mu)^2}{2\sigma^2}\right). \tag{1.9.3}$$

The pdf for the log is hardly Gaussian! This is known as a *log normal distribution*.

The formula for a change of variable is a bit more complicated if $g(x)$ is not a monotone function. See Figure 1.5.

If $g(x)$ lies within the interval y, $y + \Delta y$, then the corresponding x interval can lie at x_1 or at x_2, or at x_3, etc. We must take into account all the possible x intervals. Let $\{x_n\}$ be the set of solutions to $g(x_n) = y$.

$$p(y) = \sum_n \frac{p(x_n)}{\left|\frac{dg(x_n)}{dx}\right|}, \tag{1.9.4}$$

where the sum runs over all x_n.

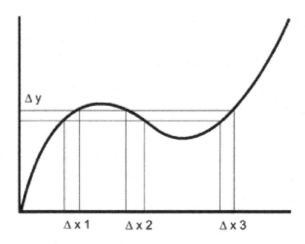

Fig. 1.5   Function with multiple roots.

Example:
Let $y = x^2$. For any y, there are two solutions $\pm x$. If the pdf for x is a Gaussian with a zero mean, the pdf for x^2 is derived as follows:

$$p(x) = \frac{1}{\sqrt{2\pi\sigma^2}} \exp\left(-\frac{x^2}{2\sigma^2}\right).$$

$$g(x) = x^2; \frac{dg}{dx} = 2x = 2\sqrt{g}.$$

Thus,

$$p(g) = \frac{2}{2\sqrt{2\pi\sigma^2 g}} \exp\left(-\frac{g}{2\sigma^2}\right) = \frac{1}{\sqrt{2\pi\sigma^2 g}} \exp\left(-\frac{g}{2\sigma^2}\right). \qquad (1.9.5)$$

Suggested Reading

1. "Statistical Design and Analysis of Engineering Experiments", Charles Lipson and Narendra J. Sheth, McGraw-Hill Book Company (New York), 1973.
2. "Probability and Statistics for Engineering and the Sciences", Second Edition, Jay L. Devore, Brooks/Cole Publishing Company, 1987.
3. "Facts from Figures", M. J. Moroney, Penguin Books, 1967.
4. "Statistical Signal Processing — Detection, Estimation, and Time Series Analysis", Louis L. Sharf, Addison-Wesley, 1991.

x1

Chapter 2

Measurement of Samples of Random Numbers

In this chapter I treat the concepts involved in dealing with measurements of random variables. If you perform a finite number of experiments, (*i.e.* take a finite number of samples of the population). the average of the results will not necessarily be $\langle x \rangle$. Indeed, the process of computing the mean of a set of numbers can be viewed as simply making a new random number by adding up N random numbers and dividing by N, *i.e.* creating a statistic. Here, we will use the concepts developed in the previous chapter to make predictions about the outcomes of real experiments.

2.1 Variance of the Measured Mean

How much can I expect a computed average to differ from the expected value? Does that result depend on how many points I use to calculate the average? Pick some number of outcomes, *i.e.* samples of a random variable, say N. Take the average of the samples. This process is called a trial. If we then do the same experiment again, with new outcomes, using the same number of samples, the new mean will usually be different from the first result. Now, consider doing the same experiment a number of times and calculating the means, being careful to not reuse any samples from one experiment to the next. The latter condition is to guarantee that the results you get are independent. The collection of averages that result from the experiments above are samples from a new random variable. What we are trying to do below is to find out how the properties of the new random variable relate to the original random variable, the outcomes.

Let $\bar{x}_n - \langle x \rangle$ be the deviation from the "true" average obtained for the nth trial of an experiment that computes the average of a random variable. What is the expected average error for that experiment or a group of experiments?

$$\langle \bar{x}_n - \langle x \rangle \rangle = \langle \bar{x}_n \rangle - \langle x \rangle \,,$$

$$\langle \bar{x}_n \rangle - \langle x \rangle = \left\langle \frac{1}{N} \sum_{k=1}^{N} x_k \right\rangle - \langle x \rangle = \frac{1}{N} \sum_{k=1}^{N} \langle x_k \rangle - \langle x \rangle \,.$$

But, by definition,

$$\langle x_k \rangle = \langle x \rangle \,.$$

Thus,

$$\langle \bar{x}_n - \langle x \rangle \rangle = 0 \,. \tag{2.1.1}$$

The "expected" error is zero. We have not proven that there is no difference between the measured average and the expected value. We have only shown that the expected value of the average is the same as the expected value of the original population. The process of taking the mean is unbiased.

The usual measure of the expected difference of the measured value from the expected value is the expected value of the square of the average error, the variance, σ_m^2.

$$\sigma_m^2 = \langle (\bar{x}_n - \langle x \rangle)^2 \rangle = \left\langle \left(\sum_{k=1}^{N} (x_k - \langle x \rangle)/N \right)^2 \right\rangle \,,$$

where k is an integer denoting the kth sample for that trial and $\bar{x}_n$ is the average obtained in the nth trial. Now, let us attempt to compute this variance. I assume that each measurement is *independent of the others*.

$$\sigma_m^2 = \left\langle \left(\sum_k (x_k - \langle x \rangle) \right) \left(\sum_p (x_p - \langle x \rangle) \right) \right\rangle \bigg/ N^2$$

$$\sigma_m^2 = \frac{\left\langle \sum_k (x_k - \langle x \rangle)^2 \right\rangle + \left\langle \sum_{k \neq p} (x_k - \langle x \rangle)(x_p - \langle x \rangle) \right\rangle}{N^2}$$

Since the points x_n are statistically independent,

$$\langle (x_k - \langle x \rangle)(x_p - \langle x \rangle) \rangle = \langle (x_k - \langle x \rangle) \rangle \langle (x_p - \langle x \rangle) \rangle \,.$$

But,

$$\langle (x_k - \langle x \rangle) \rangle = 0 \,,$$

so

$$\sigma^2_m = \frac{\left\langle \sum_k (x_k - \langle x \rangle)^2 \right\rangle}{N^2}$$

$$= \frac{\sum_k \langle (x_k^2 - 2x_k \langle x \rangle + \langle x \rangle^2) \rangle}{N^2} = \frac{\sum_k (\langle x_k \rangle^2 - \langle x \rangle^2)}{N^2}.$$

But,

$$\langle x_k^2 \rangle = \langle x^2 \rangle$$

and

$$\sigma^2 = \langle x^2 \rangle - \langle x \rangle^2,$$

the variance of the original data set, thus

$$\sigma^2_m = \frac{\sum \sigma^2}{N^2} = \sigma^2 \frac{N}{N^2} = \frac{\sigma^2}{N}. \tag{2.1.2}$$

The expected error (standard deviation) in any one measurement of the average using N points is Eq. 2.1.2.

The square of the average error for N independent samples varies proportionally to the variance of the data set and inversely as the number of points used to calculate the average. The standard deviation for N independent samples varies inversely as the square root of the number of samples used. The more points you use in calculating an average, the closer you can plan on being to the expected value.

An equivalent formula for continuous variables will be derived in a later section.

2.2 Estimate of the Variance

Now that you know the formula for estimating the variance in calculations of the mean, you may have noticed that you need an estimate of the variance of the original data set s^2. (I follow the convention used in most statistics books of using Greek characters for expected values and Roman characters for estimated quantities.) For real problems, you do not know the variance of the original data set. You can only know an estimate of the variance.

$$s^2 = \frac{\sum_{k=1}^{N} (x_k - \bar{x}_N)^2}{N - 1}. \tag{2.1.4}$$

What is the $N - 1$ about? Hang on and you will see shortly. Below, we will calculate the expected value of the estimate of the variance of the data set. Here, $\bar{x}_N$ is the computed average for the same data set using N samples.

Remember that

$$\bar{x}_N = \sum_p \frac{x_p}{N},$$

thus

$$\langle s^2 \rangle = \left\langle \left(\left(\sum_k (x_k) \right) - \left(\sum_p x_p \right) \Big/ N \right)^2 \right\rangle \Big/ (N-1)$$

$$\langle s^2 \rangle = \frac{\left\langle \sum_k (x_k^2 - 2x_k \bar{x}_N + \bar{x}_N^2) \right\rangle}{N-1}$$

$$\langle s^2 \rangle = \frac{\left\langle \sum_k \left(x_k^2 - 2x_k \frac{\sum_p x_p}{N} + \frac{\sum_p \sum_q x_p x_q}{N^2} \right) \right\rangle}{N-1}.$$

In this next step, it is important to mind your p's and q's. In other words, you need to note what happens with the counting subscripts on the variables x. Recall from calculus that these are so-called dummy variables. You can change the name of the variable without changing the result. The variable under the sum terms simply means "count all of them."

$$\langle s^2 \rangle = \frac{\left\langle \sum_k x_k^2 - 2\frac{\sum_k x_k \sum_p x_p}{N} + \frac{\sum_p x_p \sum_q x_q}{N} \right\rangle}{N-1}.$$

The N in the denominator of the last term comes about because there was no k subscript in the term, so doing the summation just multiplied the term by N. So now, we can rewrite the expression as

$$\langle s^2 \rangle = \frac{\left\langle \sum_k x_k^2 - \frac{\sum_k \sum_q x_k x_q}{N} \right\rangle}{N-1} = \frac{\sum_k \langle x_k^2 \rangle - \frac{\sum_k \sum_q \langle x_k x_q \rangle}{N}}{N-1}.$$

But $\langle x_k x_q \rangle = \langle x \rangle^2$, when $k \neq q$, and $\langle x_k x_q \rangle = \langle x^2 \rangle$, when $k = q$. In the double sum, there are $N(N-1)$ terms where $k \neq q$ and N terms where

$k = q$. Thus,

$$\langle s^2 \rangle = (N\langle x^2 \rangle - (N\langle x^2 \rangle + N(N-1)\langle x \rangle^2)/N)/(N-1)$$

$$= \frac{N-1}{N-1}(\langle x^2 \rangle - \langle x \rangle^2)$$

$$\langle s^2 \rangle = \sigma^2. \tag{2.2.2}$$

Formula 2.2.1 gives an unbiased estimate of the variance. If we had used N in the denominator, the estimate of the variance would have been biased. In some texts, it is pointed out that the original data set has N degrees of freedom (number of independent data points). You use up a degree of freedom when you calculate the average to be used in the formula for the estimated variance. The denominator of the expression is the number of degrees of freedom, $N - 1$. As an exercise, check the expected value of the following expression to see if it is biased.

$$\langle s_v^2 \rangle = \frac{\left\langle \sum_k (x_k - \langle x \rangle)^2 \right\rangle}{N}.$$

2.3 Variance of the Measured Variance

It is instructive to look at the error in the estimate of the variance. I will cheat a little in order to keep the results understandable. Let

$$\varepsilon^2 = \left\langle \left(\frac{\sum (x_n - \bar{x})^2}{N-1} - \langle \Delta x^2 \rangle \right)^2 \right\rangle,$$

the expected variance of the measured variance from the true variance, which I have written here as the second central moment

$$\varepsilon^2 \approx \left\langle \left(\frac{\sum (x_n - \langle x \rangle)^2}{N} - \langle \Delta x^2 \rangle \right)^2 \right\rangle,$$

where I have approximated $\bar{x}$ by the expected value $\langle x \rangle$. Expanding this term and assuming independent measurements, I end up with

$$\varepsilon^2 \approx \frac{1}{N}(\langle \Delta x^4 \rangle - \langle \Delta x^2 \rangle^2). \tag{2.3.1}$$

The estimated variance in the second central moment is a function of the fourth central moment. We will see other examples of this kind of behavior later. As an example, assume the pdf of $\{x_i\}$ is Gaussian. Then

$$\langle \Delta x^4 \rangle = 3 \langle \Delta x^2 \rangle^2 .$$

Thus

$$\varepsilon^2 \approx \frac{2}{N} \langle \Delta x^2 \rangle^2 . \tag{2.3.2}$$

A similar result for Gaussian variables could have been obtained using the Chi-squared distribution — see Section 1.7.7.

Exercises

Thanks to modern computers and readily available software, we can examine what the derivation just shown really means. Using the Excel® spreadsheet program, make sets of tables of random numbers uniformly distributed between 0 and 1, using the *uniform* random number generator. Make the arrays 10×100, 20×100, 50×100, 100×100. To do this in Excel®, Click "Tools," then click "Data Analysis." If you do not see "Data Analysis," you will need to load the module by clicking "Add-ins." loaded, you should scroll down until you see "Random Number Generation." Click on the button to the side of "Distribution." Select "Uniform." Parameters should be set to "between 0 and 1." To make the 20×100 set, type "20" into the "Number of Variables" box and type "100" into the "Number of Random Variables" box. Now, decide where you want to put the numbers. I usually select "Output Range" and start the set at A1. I would put each of the sets on a different worksheet ply.

Save the sets of random numbers. We will use them to illustrate many of the points discussed above.

a) Estimate 100 means and variances for each set. Clearly label and save each of the sets of calculations. They will be used again later.

i) What is the mean of the means for each set $(10 \times 100, 20 \times 100, \ldots)$?

ii) You have 100 means for each set. What is the variance of the means for each set $(10 \times 100, 20 \times 100, \ldots)$?

iii) Is there a pattern to the variance of the means as the number of points used to compute them change? What should the pattern be? Compare the theory to your measurements. Comment on whether you think the theory convincingly predicts what you observed. What else do you think you need to know to make your answers more definitive? What about the expected range of the variances of the measured variances?

When I ran the examples, I got

$$\text{VAR}_{10} = 0.00811$$
$$\text{VAR}_{20} = 0.00322$$
$$\text{VAR}_{50} = 0.00149$$
$$\text{VAR}_{100} = 0.00067$$

What would we have predicted as the expected value for these measurements? The Uniform distribution $[0,1]$ was used. The expected mean is given by

$$\langle x \rangle = \int_0^1 x\,dx = \left.\frac{x^2}{2}\right|_0^1 = \frac{1}{2}\,.$$

The expected variance for the uniform distribution is given by

$$\sigma^2 = \int_0^1 (x-0.5)^2 dx = \left.\frac{(x-0.5)^3}{3}\right|_0^1 = \frac{(0.5)^3 - (-0.5)^3}{3} = \frac{1}{12} \approx 0.08333\,.$$

According to Equation 2.1.2, the expected value of the measured variance of the means should be given by

$$\langle \text{VAR}_{10} \rangle = 0.08333/10 = 0.00833$$
$$\langle \text{VAR}_{20} \rangle = 0.08333/20 = 0.00416$$
$$\langle \text{VAR}_{50} \rangle = 0.08333/50 = 0.00167$$
$$\langle \text{VAR}_{100} \rangle = 0.08333/100 = 0.000833$$

You can see from the figure that the trend in the variance of the mean is roughly what it should be. However, since a measured variance is a statistic, we do not expect an exact match between the expected result and the measurement. However, we can look to see if the measured result is as close to the expected result as we would expect. We can do the check by estimating the *expected variance of the measured variance*.

By 2.3.2, the expected variance of the measured variance should be approximately

$$\varepsilon^2 \approx \frac{2}{N}\langle \Delta x^2 \rangle^2\,.$$

In this case, $\langle \Delta x^2 \rangle$ is the expected variance. The value of N is the number of points used to calculate the variance, here 100 points. The measure we are looking for is the expected standard deviation. If the difference between the expected variance and the measured variance is on the same order of magnitude as the expected standard deviation of the measured variance, we can be somewhat confident that Equation 2.1.2 is governing the magnitude of the measured variance of the means. To be a bit more precise, if the pdf

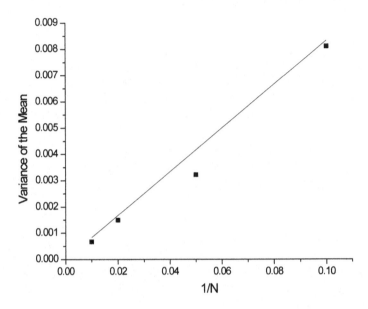

Fig. 2.1   Plot of the variance of the mean vs. 1/N. The line is the theory and the points are the measurements.

of the measured variances were a Gaussian, we would expect to find 95% of the measured variances between

$$\sigma_N^2 \pm 1.96 \times (\text{standard deviation of the variance}).$$

A lot of people just use 2 instead of 1.96. Here, the standard deviation is approximately given by

$$\sigma_{s^2} \approx \sqrt{\frac{2}{N_s} \frac{\sigma^2}{N}},$$

where σ^2 is the variance of the original data set, N_s is the number of points used to estimate s^2, and N is the number of points used to estimate the average. (Please read over the previous couple of sentences very carefully. There are many N's being thrown around and you can become completely confused if you do not understand why which N is being used.) For instance, the estimated variance for means obtained using 10 samples has $N = 10$, and $N_s = 100$.

For the expected standard deviation for the variance of the averages obtained using 10 points I get

$$\sigma_{s^2} \approx \sqrt{\frac{2}{100}}(0.0833/10) = 0.001178\,.$$

In summary, the expected measurements, to a confidence of 95% are:

$$\langle \text{VAR}_{10} \rangle = 0.00833 \pm 2 \times 0.001178$$
$$\langle \text{VAR}_{20} \rangle = 0.00416 \pm 2 \times 0.000589$$
$$\langle \text{VAR}_{50} \rangle = 0.00167 \pm 2 \times 0.000236$$
$$\langle \text{VAR}_{100} \rangle = 0.000833 \pm 2 \times 0.000118$$

The theoretical variances of the means all fit within the 95% confidence limits of the measured variances, giving some confidence that our measurements support 2.1.2. Despite what you read in the papers and see on TV, you cannot use statistics to prove things. You can show that your explanation of what is going on is likely to be true, but, again, you cannot prove it.

Exercise:

Make a histogram of the means for each set made in the previous task. See Section 2.5 for details on making a histogram.

a) Does the distribution of the means have the same shape as the original distribution?

b) Describe the Central Limit Theorem. What does it have to do with your results in a)?

c) Using the 20×100 data set, calculate the estimated variance of each set of 20 numbers. Then, replot the distribution of the means as a t-distribution.

d) Explain how to pick the interval that will likely contain 97% of the outcomes.

2.4 Non-Independent Random Variables

In almost everything above, we have assumed independent random variables. What happens if the random variables are not independent? As a practical matter, here, non-independent means that the fluctuation of the mth point from the expected value correlates with the fluctuation from the expected value of the nth for some values of m and n. The average is still an unbiased estimate of the expected value (see Equation 2.1.1). However, the evaluation of the variance of the mean changes. Again,

$$\sigma_m^2 = \left\langle (\bar{x}_m - \langle x \rangle)^2 \right\rangle = \left\langle \sum_k (x_{km} - \langle x \rangle)^2 / N^2 \right\rangle$$

$$\sigma_m^2 = \left\langle \left(\sum_k x_{km} - \langle x \rangle \right)^2 \right\rangle \bigg/ N^2$$

where m and n are integers and $\bar{x}_n$ is the average of the nth experiment. x_{km} is the kth element of the data set used to calculate $\bar{x}_m$. I again assume that there are N total measurements in each experiment. This time, however, we do not assume the measurements are independent.

$$\sigma_m^2 = \frac{\sum_k \langle (x_{km} - \langle x \rangle)^2 \rangle + \sum_{p \neq k} \sum \langle (x_{pm} - \langle x \rangle)(x_{km} - \langle x \rangle) \rangle}{N^2}$$

Now define

$$R_{nm} = \langle (x_n - \langle x \rangle)(x_m - \langle x \rangle) \rangle / \sigma^2, \qquad (2.4.1)$$

the nm covariance coefficient.

Further define the average covariance, R, by

$$R = \frac{1}{N(N-1)} \sum \sum_{n \neq m} R_{nm}. \qquad (2.4.2)$$

Then, it is easy to show

$$\sigma_m^2 = \frac{\sigma^2}{N} (1 + (N-1)R). \qquad (2.4.3)$$

The variance of the mean is strongly affected by the degree of covariance in the data set. In a sense, the number of independent data points in the data set is $N/(1 + (N-1)R)$. The more the interdependence of the data shown by the covariance, the fewer the number of independent data points in the data set. The parameter R is 0, when there is no correlation in the data set, and 1, when the data set is completely correlated. In general, it will be between 0 and 1. Note also that the value of R depends on the number of data points used in computing the mean values. There will be more on calculations of means with correlated data in the next chapter.

2.5 Histograms

A histogram is a method of representing the outcomes of a set of samples of random variables. It is a set of numbers showing the frequency of appearance of a random variable in a given state. If the random variable takes on discrete values, $i = \{1, 2, 3, \dots\}$ the histogram represents the number of times the state i appeared in a set of trials. If the random variable has a continuous distribution, the histogram represents the number of times the variable appeared in some interval, for instance, $[x_i - \delta x/2, x_i + \delta x/2]$.

For a discrete random variable, the interpretation of the histogram is straightforward. In the limit that the number of trials goes to infinity, the histogram divided by the total number of trials goes to the pdf for the variable. For a continuous variable, you have to be a bit more careful. Let h_i be the expected value of the ith value of a histogram for a continuous variable x. Then

$$\langle h_i \rangle = N_T \int_{x_i - \delta x/2}^{x_i + \delta x/2} p(x)dx \,,$$

where N_T is the total number of measurements and $p(\ )$ is the pdf for x. For ease of interpretation of the experiment, you usually want to approximate the integral above by

$$\int_{x_i - \delta x/2}^{x_i + \delta x/2} p(x)dx \approx p(x_i)\delta x \,.$$

It can be shown that meeting this condition requires that

$$\delta x \ll \left(\left| \frac{3f(x_i)}{f''(x_i)} \right| \right)^{1/2} . \tag{2.5.1}$$

The second derivative of a pdf is usually at a maximum near its peak, so the curvature near that point determines the maximum width, δx. On the other hand, if δx is made too small, you have the possibility of getting too few samples in the interval to be useful. There are several software packages that have good heuristic schemes for picking δx. I recommend that you pick your own value, examine the result and alter δx until you are satisfied that you have a useful set of intervals.

For instance, if you think you have a Gaussian distribution, at the center where the second derivative is maximum, the criterion above gives

$$\delta x < \sqrt{3}\sigma \,.$$

Recall that the statistics of the measurements in the histogram are Binomial. You will need to take that into account if you are trying to do a curve fit to a histogram. There will be more details on this process in Chapter 4.

Exercise:

Set up a generator for a uniform distribution between (0,1) and a Gaussian generator that has zero mean and a variance of one.

a) Run both generators and estimate the variance and the fourth moment of each of them. How are the variance and the fourth moments related?

b) Using the Gaussian Noise Generator, make 400 random numbers. Make a histogram of the numbers. Is the generator really Gaussian? Can you tell? Estimate the mean and the variance of the data set using the histogram and separately using the entire data set. If you are really curious, widen the intervals in the histogram and repeat the exercise above until there is a significant difference between the results of the two methods. What causes the difference?

2.5.1 *Statistical experiments involving the uniform and binomial distributions*

Open a new Excel® Spreadsheet. First create a set of 1000 random numbers from a uniform distribution. I did it by calling up Data Analysis under "Tools" on the Toolbar. Make 1000 numbers between 0 and 1, by setting the number of variables to 10 and the number of random numbers to 100. The default range for the uniform generator is [0,1]. If you want to get fancy, you can change this range, but it is unnecessary for this exercise. Now, make a histogram of the data, using 100 bins. In other words, you want to divide the interval 0,1 into bins of length 0.01. I made a column of 100 numbers starting with 0.0, 0.01, 0.02, ... up to 1.0. Then, I called up "histogram" and made a histogram of the 1000 numbers using the bin range described above.

My histogram is shown in Figure 2.2.

Note that although the histogram we made looks vaguely flat, it actually has a lot of variation in it ranging from about 5 counts in a bin to around 20 counts in a bin. By definition, the histogram should average 10 counts per bin, since there are 100 bins to hold 1000 counts. The key to understanding the distribution of counts in the bins is the binomial distribution. This is a

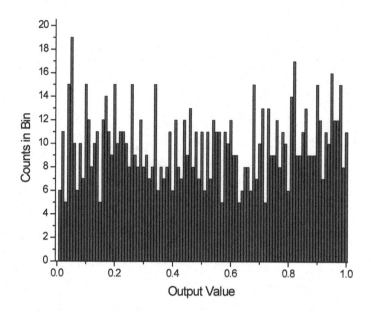

Fig. 2.2   Histogram of uniform distribution sample.

classic case where we have an expectation of 10 counts in a bin with 1000 trials. To be formal, the probability for each bin is 0.01 since there are 100 bins and each one is as likely to get a measurement as any other for each trial (new random number). There are a total of 1000 trials, so using formula 1.6.6, we can calculate the probability of our observations. We can estimate how often we should see 5 counts or 10 counts, etc.

In terms of making the histogram of the number of measurements in each bin, the total number of example counts is 100 — the number of trials (bins). I make the histogram shown in Fig. 2.3 by counting the number bins with 1 count, 2 counts, 3 counts, etc. So, if I calculate the probability of 10 measurements out of 1000 trials in a bin with a probability or 0.01 using Equation 1.7.4, I get $p = 0.12574$. Out of 100 results, the expected number of times that I should see 10 counts in a bin is $100 \times 0.12574 = 12.574$. In the trial I ran, I got 10 counts 9 times. The complete histogram for my example is shown below. The bars are the data and the smooth curve is the theoretical result calculated using the binomial formula.

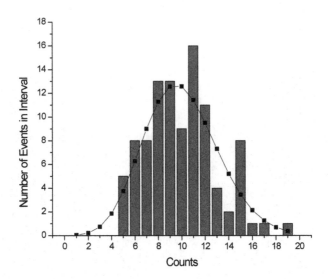

Fig. 2.3   Measured histogram of the histogram of the uniform random number generator. The smooth line is the expected number of measurements if the distribution is binomial.

Does this picture make sense to you? Is the variation of each bin from the expected value within the range of what could be expected? What more do you need to know to be able to answer these questions confidently?

In order to come up with quantitative answers to the questions above, we use the binomial distribution. What we will be doing is essentially calculating the probability of the data set that we are dealing with. Later in these notes, the concept of the probability of a data set will be central, so pay attention to how it is used here.

At the peak of the distribution, I compute the expected number of counts to be 12.574. The measured number of counts is 9. By Equation 1.7.4, the probability of getting 9 counts when the expected number is 12.574,

$$p(9) = \binom{100}{9} \mu^9 (1-\mu)^{91} = \binom{100}{9} (0.1274)^9 (0.8726)^{91} = 0.075 \, ,$$

about 1 in 14. The probability of getting 12 counts, which is closer to the expected value is 0.117. Getting 9 counts is hardly a rare event.

2.6 Confidence Limits

A question that often arises is, "If I take a sample of a random variable with a known pdf, what is the chance that the sample will be larger than some value, say x_m?" Alternatively, you could ask what are the chances that the measurement is between some limits x_l and x_u? The answer to these questions is easily obtained.

Equation 1.3.9 gives the answer to the first question.

$$p(x \rangle x_m) = 1 - \int_{\text{min value of } x}^{x_u} p(x')dx'$$

$$= \int_{x_l}^{\text{max value of } x} p(x')dx' . \qquad (2.6.1)$$

The answer to the second question is likewise easy to derive.

$$p(x_l \le x \le x_u) = \int_{x_l}^{x_u} p(x')dx' . \qquad (2.6.2)$$

These answers assume that you know the pdf for the random variable being sampled.

A related question that has a slightly more satisfying answer is, "What are the chances, if I measure the mean of a random variable by using N samples, that the measured mean will be within some specified range of the expected value?" Another way to ask the same question is to ask what the confidence limits are on the measurement of the mean. This question is easier to answer, because the mean is, in fact, the sum of N random variables, thus the pdf of the mean is approaching being a Gaussian with a variance of σ^2/N (See 1.7.17).

The procedure for addressing the confidence limits is as follows:
Make a new random variable

$$z = \frac{(x_m - \mu)\sqrt{N}}{\sigma} , \qquad (2.6.3)$$

where μ is the expected ("true") mean of the data set, x_m is the measured mean, and σ is the standard deviation of the data set. By the Central Limit Theorem, if there are enough terms used to make the average, this new variable z will have a Gaussian pdf with a mean of zero and a variance of one.

The chance that this new variable z will be between limits $\pm z_{a/2}$ is thus given by

$$p(-z_{a/2} \leq z \leq z_{a/2}) = \frac{1}{\sqrt{2\pi}} \int_{-z_{\alpha/2}}^{z_{\alpha/2}} \exp\left(-\frac{x^2}{2}\right) dx$$

$$= \operatorname{erf}\left(\frac{z_{a/2}}{\sqrt{2}}\right). \tag{2.6.4}$$

The z term is usually picked by the desired probability. If, for instance, you want to guarantee that the two-sided probability is 95%, then the area above $z_{a/2}$ and below $-z_{a/2}$ should be 0.025. So, here, $a/2$ is 0.025.

The chances that a measured mean will give a z outside this limit is given by

$$Prob(z \text{ outside the limits } \pm z_{a/2}) = 1 - \operatorname{erf}\left(\frac{z_{a/2}}{\sqrt{2}}\right). \tag{2.6.5}$$

To find the 95% confidence limits, for instance, you simply have to find the value of $z_{a/2}$ for which Equation 2.6.4 equals 0.95 and then solve for x_m. By using a simple search routine on the formula for the error function, you would find that the appropriate $z_{a/2}$ value for 95% would be 1.96.

$$z_{a/2} \geq \left|\frac{(\bar{x} - \mu)}{\sigma/\sqrt{N}}\right| \text{ or } z_{a/2}(\sigma/\sqrt{N}) \geq |(\bar{x} - \mu)| . \tag{2.6.6}$$

For example, if you knew you were dealing with Gaussian random variables and had a measured variance of 4 with 20 measurements,

$$z_{a/2} = 1.96 \geq \left|\frac{(\bar{x} - \mu)\sqrt{20}}{2}\right| \text{ or } 0.877 \geq |\bar{x} - \mu| .$$

In other words, you can claim that 95% of the time, you measured the correct mean $\pm$ 0.877.

Of course, one can do one-sided versions of the calculations shown above.

There was a bit of a fudge in the discussion above. In reality, you only have estimates of the variance of the data set and thus only an estimate of the variance of the mean. To do this problem correctly, you would have to take that problem into account and use a Student-t distribution for the pdf of the means. For a lot of practical problems, the difference is negligible. For the Student-t case you would have

$$t_{a/2,N-1} \geq \left|\frac{(\bar{x} - \mu)}{s/\sqrt{N}}\right| \text{ or } t_{a/2,N-1}(s/\sqrt{N}) \geq |(\bar{x} - \mu)| . \tag{2.6.7}$$

The equivalent t value for the 95% confidence limits for 19 points is 2.093, not much different from the 1.96 found for Gaussian variables. Many computer packages, including Excel® have software that will give the desired t values.

2.6.1 Tests of hypotheses

A question that often comes up is whether a change in a procedure results in a real change in the output. Examples range from changing the amount of fluorine in drinking water to affect the rate of tooth decay in the general populace to adjusting a tool feed to affect the diameter of a crankshaft. What complicates life is that there is usually a natural variation in the output with *no* change in procedure. What you need to determine is whether the change in procedure made a difference in the output and, if you can figure it out, how much of a change?

As you might imagine, the answer can only be given in a probabilistic sense. You will not be able to say "I am certain that the new procedure is better, worse, or just the same!" (pick one). The key to all of what follows is analysis of the change in order to ascertain what change could be expected by chance compared to the change actually measured. If the measured change is outside of what you could usually expect by chance, you are allowed to proclaim the change "significant."

There are basically two approaches to this problem: 1) Pick the size of a significant variation and then compare actual results to this targeted level. 2) Make the measurements and then attempt to interpret the probability of change vs. the probability of no change.

A formal language has grown up around this problem. The *null hypothesis* is the assumption that there is no change in the expected values. A type 1 error is the designation for the situation where the null hypothesis is true, but the data leads the experimenter to reject the hypothesis. A type 2 error is accepting the null hypothesis when it is false.

Imagine that initially the output was a random variable with an expected value of μ_1 and a variance of σ_1^2. After the change, the expected value is μ_2 and the variance is σ_2^2. If you knew this information, you would not have a problem. But, the point is you usually don't know the expected values and variances, so how do you proceed?

Let's look at the easy problem first. Say we have an expected value μ and have picked a positive level of significant change, $\pm\Delta\mu/2$. This is the size of the change that is important to you. The question that needs to be answered is whether you can reliably tell if such a change has taken place. What you want is when $|\bar{x} - \mu| \geq \Delta\mu/2$, you can claim that there was a significant change in the expected value of the output. What is the probability that you are correct in claiming such a change? What is the probability that you thought you had a significant change in the system when there was not a change?

In the case that you know the variance of the measurements, σ^2, you can create the variable

$$z = \frac{\bar{x} - \mu}{\sigma/\sqrt{N}}.$$

I assume that $\bar{x}$ is a Gaussian variable. This is likely because the mean is the weighted sum of a bunch of random variables and the central limit theorem tells us this new random variable will tend to be a Gaussian. Then z is also a Gaussian variable with a mean of zero and a variance of one. Thus, the probability of a random measurement being larger than $\Delta\mu$ is the same as the probability of

$$z \geq \left|\frac{\Delta\mu/2}{\sigma/\sqrt{N}}\right|.$$

We know the answer to this question from previous work, namely Section 2.6.1 and Equation 2.6.6.

$$p\left(z \geq \left|\frac{\Delta\mu/2}{\sigma/\sqrt{N}}\right|\right) = 1 - \text{erf}\left(\left|\frac{\Delta\mu/2}{\sigma/\sqrt{N}}\right|\Big/\sqrt{2}\right). \qquad (2.6.8)$$

Formally, the above expression is the probability of a type 1 error, finding that the null hypothesis is false, when it is really true. It is the probability that a random fluctuation is bigger than the chosen significant level, even when nothing has changed in the system.

The probability of a type 2 error is likewise straightforward to compute. Let the original expected value be μ_1 before the change and μ_2 after the change. If you keep the same $\Delta\mu$ as above, the probability for a type 2 error, i.e. the chance the output z value will be between $-\Delta\mu/2$ and $\Delta\mu/2$ is given by

$$\frac{1}{\sqrt{2\pi}} \int_{-\frac{\Delta\mu\sqrt{N}}{\sigma}}^{\frac{\Delta\mu\sqrt{N}}{\sigma}} \exp\left(-\left(z - \frac{(\mu_2 - \mu_1)\sqrt{N}}{\sigma}\right)^2\Big/2\right) dz. \qquad (2.6.9)$$

This expression can be put in a more easily computed form, *viz.*

$$\left| \frac{\mathrm{erf}\left(\dfrac{(\mu_2 - \mu_1 + \Delta\mu/2)\sqrt{N}}{\sigma\sqrt{2}}\right) - \mathrm{erf}\left(\dfrac{(\mu_2 - \mu_1 - \Delta\mu/2)\sqrt{N}}{\sigma\sqrt{2}}\right)}{2} \right| . \quad (2.6.10)$$

Unsurprisingly, the probability of missing a change when there is one depends on how large the change is compared to the natural variation in the measured means.

To do a numerical example of the case we just looked at, let the variance be 4 and again assume 20 measurements were used to measure the mean. The expected value is 10.5 and $\Delta\mu/2 = 0.7$. Putting these numbers into 2.6.9, we find that the probability of a type 1 error is 0.118, a bit more than 1 out every 9 measurements when there is no change, will be interpreted as a change.

The probability for a type 2 error is a function of the change in the expected value, the variance, as well as the number of points used to calculate the mean. Using the numbers above and letting the change in the expected value be δ, you can make a plot of the probability of a type 2 error as a function of the change in expected value.

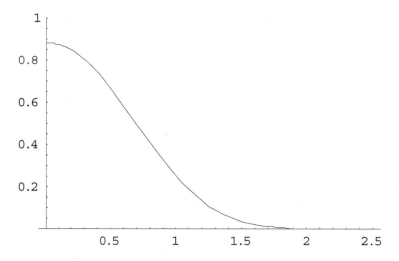

Fig. 2.4 Probability of type 2 error as function of separation between expected values, δ, for a sample size of 20 points.

You can see that when the change in the expected value is 0.7, the chance of a type 2 error is about 0.5. You will get it wrong about half the time. On the other hand, when the change is 2.0, the chance of getting the wrong result is negligible. If you want more certainty about a change around 0.7, then you need to use more data points to calculate the mean. Here, instead of 20 points, 40 points would be better in that you would make far fewer type 2 errors.

A concern is how you report on the result of an experiment like the one above. One possibility is to report the t-value (z-value) that you used for the critical value and whether the measurement exceeded it. Another possibility is to report the actual t-value or the probability of a type 1 error for that t value. This latter number is called the p-value. It is the probability of a type 1 error given your measurement.

Now, let us look at a somewhat messier problem. If you take N measurements of the output before you make the change, you will have an estimate of the mean $\bar{x}_1$ and an estimate of the variance of the process s_1^2. Likewise, you have M measurements of the output after the change, yielding $\bar{x}_2$ and s_2^2. How do you decide if there has been a change in this case?

If the null hypothesis is true, the appropriate distribution for the difference of the measured means should be a Student-t distribution with an expected value of zero, $N + M - 2$ degrees of freedom[1], and a variance of

$$\left(\frac{s_1^2}{N} + \frac{s_2^2}{M} \right)$$

$$t_{N+M-2} = \frac{(\bar{x}_1 - \bar{x}_2)}{\sqrt{\left(\frac{s_1^2}{N} + \frac{s_2^2}{M} \right)}} \, .$$

Then you proceed exactly as above. You pick a critical level and make your decision based on whether the measurement showed a change bigger than the critical level. You could also report the p-value.

For example, if you wish to keep the type 1 error rate under 5% while using 20 points for each average, you will accept the null hypothesis for t-values that are between -2.02 and 2.02. Most texts write this as

$$t_{0.025, 38} = 2.02 \, .$$

---

[1]Strictly speaking, this is only true when the expected variance is the same for both. My experience is that the penalty for making this assumption when the variances are not equal is trivial.

To make this more concrete, let us assume that we have a sample measurement set with $N = 10$, $M = 12$, yielding

$$\bar{x}_1 = 2.3$$

and

$$\bar{x}_2 = 2.75,$$

with

$$s_1^2 = 0.51 \text{ and } s_2^2 = 0.29.$$

Then

$$t^2 = \frac{.45^2}{\left(\dfrac{0.51}{10} + \dfrac{0.29}{12}\right)} = 2.69, \qquad t = 1.64,$$

$$p(t = 1.64) = 0.117.$$

Then, to a level of significance of 11.7%, the null hypothesis is rejected. The p-value is 0.117. More plainly, you have a chance of 11.7% of getting a difference at least that big in the measured means, even if there was no difference in the true means.

Exercise: The exercise covers comparison of population statistics using the standard t-test.

Using the Normal Random number generator in Excel®, generate two sets of 1200 numbers (10×120). The first should have an expected value of 1500 and a standard deviation of 495. The second set should have an expected value of 1800 and a standard deviation of 595. Your job is to examine small samples of each set to tell whether the two sets are from the same population.

a) Estimate the mean and variance of 10 pts from each set. Can you tell whether the means of the two populations are different? To what level of confidence?

b) Do the same thing 4 more times, using a different set of points each time. Do you get the same result as above?

c) Generate 100 examples of comparisons between the two populations. Combine your results to make a single histogram.

i) What does this histogram tell you about your earlier result in a) and b)?

ii) Pick a tolerance level and a desired level of confidence and then estimate the number of samples you would need from each population to achieve these levels in predicting the difference between the populations.

iii) You also had to make some assumptions about the variances of the populations. What were they? What practical difference did the assumptions make? Excel® gives three different types of t-tests. Try all three and see whether there is any practical difference between them.

Exercise: This exercise exams various pdfs concerning the estimated variance.

Open a new Excel® workbook and create two sets of random numbers. The first set should be a 5×100 set of "normal" distribution with a mean of 0 and a variance of 8. The second set should be 10×100 and be a normal distribution with a mean of 10 and a variance of 14.

a) Calculate the variance for each row of 5 points and 10 points respectively.

i) Using one row (5 points) of the first set, calculate the 95% confidence interval for the expected variance. Redo using 4 more rows. Do these results correspond to your input? Do again using the set having 10 points per row.

b) Calculate a histogram of the variances for each set.

i) Rescale each histogram and compare to the appropriate χ^2 distribution. First, simply compare the mean and variance of the data to the expected values computed from the formulae for the χ^2 distribution. Then, compare the measured distributions to the predicted distributions by making a plot showing the measured value and the theoretical value for the histogram intervals.

Suggested Reading

1. "Statistical Design and Analysis of Engineering Experiments", Charles Lipson and Narendra J. Sheth, McGraw-Hill Book Company (New York), 1973.
2. "Probability and Statistics for Engineering and the Sciences", Second Edition, Jay L. Devore, Brooks/Cole Publishing Company, 1987.
3. "Facts from Figures", M. J. Moroney, Penguin Books, 1967.
4. "Probability, Random Variables, and Stochastic Processes", A. Papoulis, McGraw-Hill Book Company, 1965.

Chapter 3

Time Series-Random Variables that are Functions of Time

3.1 Averages

Let $f(t)$ be a fluctuating function of time. It is a random variable. We shall follow the convention usually used in electrical engineering and call this a random signal. We define a time average of f as follows:

$$\bar{f}_T = \frac{1}{T} \int_0^T f(t) \, dt \tag{3.1.1}$$

$$\bar{f} = \lim_{T \to \infty} \bar{f}_T \,. \tag{3.1.2}$$

The result, $\bar{f}_T$, is an estimate of the time average $\bar{f}$. A useful theorem expresses a relation between the time average and the ensemble average from the previous section: the Ergodic Theorem. It states, "If $\bar{f}$ and $\langle f \rangle$ both exist, $\langle f \rangle = \bar{f}$." The time average is the expected value.

In principle, the time average estimate can depend on when the averaging is performed. For instance, if you had a large vertical water tank with a valve at the bottom, the average velocity measured at the valve opening would depend on how long the valve had been open (unless the tank were continuously refilled). The average velocity would decrease as the tank emptied. Another example would be the average windspeed during a thunderstorm. At the peak of the storm the average wind speed measured is likely to be higher than the average measured at the start or at the beginning of the storm. However, here we will consider only functions where the average is expected to be independent of the time the measurement is done. Such processes are called *stationary* (in the strict sense). A good example of a stationary process is the turbulent flow of water in a pipe, where the external parameters such as the pressure difference, temperature, etc. have been maintained steady. The expected value of the average velocity at each point should be the same no matter when you measure it.

A *realization* of a time series is a connected, contiguous, finite subset of the infinite time series. The expected value of any function of a *stationary time series* is the same for every realization. In general this means a system whose statistics are stable with time.

For a stationary process,

$$\langle \bar{f}_T \rangle = \bar{f}. \tag{3.1.3}$$

3.2   The Autocovariance and Autocorrelation Function

In general, random variables that are functions of time contain some degree of correlation among the values of the output. Define the autocovariance of the data

$$R_{ff}(t, t+\tau) = \langle (f(t) - \langle f \rangle)(f(t+\tau) - \langle f \rangle) \rangle / \sigma^2. \tag{3.2.1}$$

If the fluctuations are slow on the time scale of τ, then $f(t)$ will be related to $f(t+\tau)$. For large τ, $f(t)$ and $f(t+\tau)$ will not be related. We will clarify the meaning of "short" and "long" later.

The definition of the autocovariance can be written

$$R_{ff}(\tau) = \lim_{T \to \infty} \frac{1}{\sigma^2 T} \int_0^T (f(t) - \langle f \rangle)(f(t+\tau) - \langle f \rangle) dt.$$

Written this way, it is can be seen that the expected covariance is only a function of the separation, τ. This should be clear because we are dealing with stationary signals. The expected value of the autocovariance should not be a function of when the expectation is taken.

$$R_{ff}(\tau) = \langle (f(0) - \langle f \rangle)(f(0+\tau) - \langle f \rangle) \rangle / \sigma^2. \tag{3.2.2}$$

The related quantity

$$RC_{ff}(\tau) = \langle f(0)f(\tau) \rangle \tag{3.2.3}$$

is called the autocorrelation function. It is related to the autocovariance by

$$RC_{ff}(\tau) = \sigma^2 R_{ff}(\tau) + \langle f \rangle^2. \tag{3.2.4}$$

If there is no ambiguity, the subscripts "*ff*" are omitted. This nomenclature is not universal, but I thought the distinction between the various "correlations" was important to note.

3.2.1 *Variance of the average of a signal with correlation.*

This arises when you are trying to measure the average of a fluctuating, correlated signal like the velocity in a turbulent flow. Once again, the process of taking the average is an algorithm for making a new random variable. Note, also, that we are dealing with a continuous signal, so the concept of an "independent estimate" needs to be defined for this particular case. In order to calculate the variance of the average, we perform similar calculations to those used in the previous chapter.

$$\sigma_m^2 = \langle (f_T - \langle f \rangle)^2 \rangle$$

$$= \langle f_T^2 \rangle - \langle f \rangle^2$$

$$= \frac{1}{T^2} \left\langle \int_0^T \int_0^T f(t') f(t'') dt' d'' \right\rangle - \langle f \rangle^2$$

$$= \frac{1}{T^2} \int_0^T \int_0^T \langle f(t') f(t'') \rangle dt' dt'' - \langle f \rangle^2 .$$

But the term in brackets is the expected autocorrelation function, so

$$\sigma_T^2 = \frac{1}{T^2} \int_0^T \int_0^T (\sigma^2 R(t' - t'') + \langle f \rangle^2) dt' dt'' - \langle f \rangle^2 .$$

$$\sigma_T^2 = \frac{\sigma^2}{T^2} \int_0^T \int_0^T (R(t' - t'') dt' dt'' .$$

Now, by changing the integration variables from t' and t'' to t' and $\tau = t' - t''$, we get

$$\sigma_T^2 = \frac{\sigma^2}{T} \int_{-T}^T \left(1 - \frac{|\tau|}{T} \right) R(\tau) d\tau . \tag{3.2.5}$$

In order to understand this expression better, we need to know more about the behavior of the autocovariance. Figure 3.1 shows a typical snapshot of a realization (time interval) of a randomly varying time signal. The solid horizontal line is the mean of the data. The autocovariance is proportional to the average value of the product of the deviation of $f(t)$ from the mean times the deviation from the mean at time $t + \tau$. This product is typically called the lag product. If τ is a "short" time interval, say one or two lag times, the lag product will overwhelmingly be a positive number. It will not deviate much from the variance of the data, σ^2. On the other hand, if the lag time is long, say 50, the lag products will tend to be equally positive and negative. The average will thus be close to zero. This is a fancy way of saying that at short times, the variable is still correlated with itself, and at long lag times, knowing the value at the beginning of the interval will not be helpful in predicting the value at the end of the interval. A typical autocovariance for data such as that in 3.1 is shown in Figure 3.2.

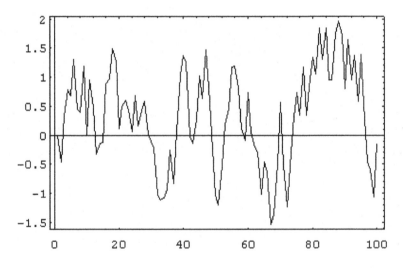

Fig. 3.1   Random time signal.

The autocovariance starts at one and decays toward zero as τ goes to infinity. Perhaps it is not obvious from the drawing, but it is easy to show that the autocovariance is always symmetric in the variable τ, *i.e.*, $R(\tau) = R(-\tau)$. This is simply another manifestation of the fact that the expected value of the lag product is only a function of the separation time.

We are now in a position to quantify the concepts of "short" and "long"

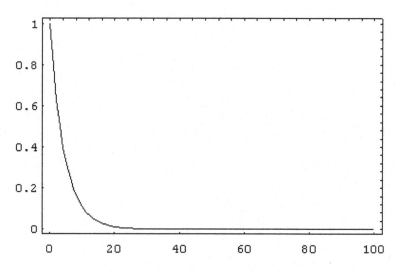

Fig. 3.2   Typical autocovariance.

lag times. The integral correlation time of a signal, Λ, is defined by

$$\Lambda \equiv \int_0^\infty R(t)dt \tag{3.2.6}$$

In a sense, this is the natural time scale of the process that makes the signal $f(t)$. In turbulent flow, this quantity is known as the "macro-scale". Long and short lag times are now understood to mean long and short compared to Λ.

Some authors, especially those concerned with turbulent flow define another natural time scale, λ, known as the micro time scale of the signal. It is defined by expanding the autocovariance in a Taylor's series about the origin ...

$$R(\tau) \approx 1 - \frac{\tau^2}{2\lambda^2} + \cdots \tag{3.2.7}$$

or

$$\lambda = (-R''(0))^{-1/2} . \tag{3.2.7a}$$

Now we are in a position to make some sense out of Equation 3.2.5, the expression for the variance of the mean of the signal measured over a time interval T. If T is larger than Λ, then

$$\sigma_T^2 = \frac{\sigma^2}{T} \int_{-T}^{T} \left(1 - \frac{|t|}{T}\right) R(t) dt \approx \frac{2\sigma^2 \Lambda}{T}. \qquad (3.2.8)$$

Examining our expressions for the variance of the mean derived above for independent data and comparing them to the expression just derived, we can state that the number of independent estimates of the mean in the time interval T is $T/2\Lambda$. The relation between the number of independent measurements and the correlation time of the signal is a very useful concept to keep in mind in the measurement of any random variable.

3.3   The Power Spectrum of a Random Signal

Define the complex spectrum of a realization of a random signal, $s(\omega)$ by

$$s_T(\omega) \equiv \frac{1}{\sqrt{T}} \int_0^T f(t) e^{-i\omega t} dt. \qquad (3.3.1)$$

The *power spectrum*, $S_T(\omega)$, is defined as

$$S_T(\omega) \equiv s_T(\omega) s_T^*(\omega) \equiv |s_T(\omega)|^2, \qquad (3.3.2)$$

where $s_T^*( )$ is the complex conjugate of $s_T( )$. The power spectrum is non-negative and is usually interpreted as a measure of the average energy present in the signal $f(t)$ at the frequency ω.

If we compute the expected value of the complex spectrum and the power spectrum, we get

$$\langle s_T(\omega) \rangle = \frac{1}{\sqrt{T}} \int_0^T \langle f(t) \rangle e^{-i\omega t} dt$$

$$= \frac{1}{\sqrt{T}} \int_0^T \langle f(t) \rangle e^{-i\omega t} dt = \langle f \rangle \frac{1}{\sqrt{T}} \int_0^T e^{-i\omega t} dt. \qquad (3.3.3)$$

The expected value of the complex spectrum contains no information about the fluctuating behavior of the signal!

$$\langle S_T(\omega) \rangle = \frac{1}{T} \int_0^T \int_0^T \langle f(t) (f(t') \rangle e^{i\omega t} e^{-i\omega t'} dt dt'.$$

Note that the quantity inside the expected value is $RC(t - t')$.

Now, pulling the, by now usual trick, of changing integration variables, we get

$$\langle S_T(\omega) \rangle = \int_{-T}^{T} \left(1 - \frac{|\tau|}{T}\right) RC(\tau) e^{-i\omega\tau} d\tau. \tag{3.3.4}$$

We have derived a version of the well-known Weiner–Khinchin theorem relating the autocorrelation function and the power spectrum. The spectrum is the Fourier Transform of the autocorrelation.

The form shown here differs from what most authors give in two aspects. First, there is the matter of the measurement time T that I have assumed to be the length of the data set from which the covariance and the spectrum are computed. Taking that into account causes the factor $(1 - |\tau|/T)$ to appear in the computations. In principle, it distorts the computation of the autocorrelation and the spectrum; however, if, $T \gg \Lambda$ the effect can be shown to be negligible. Later it will be shown that this factor can be an essential part of measuring a smooth spectrum.

3.3.1 *Short review of Fourier transforms*

The Fourier Transform of a function $f(t)$ is defined as

$$F(\omega) = \int_{-\infty}^{\infty} f(t) e^{-i\omega t} dt. \tag{3.3.5}$$

If $f(t)$ is a time domain signal, then $F(2\pi f = \omega)$ is called the frequency domain signal.

This function of ω is, in general, complex. However, it can be shown that if the function $f(t)$ is real and symmetric, the Fourier Transform is real.

$$F^*(\omega) = \int_{-\infty}^{\infty} f(t) e^{i\omega t} dt.$$

Let $x = -t$.

$$F^*(\omega) = -\int_{\infty}^{-\infty} f(-x) e^{-i\omega x} dx = \int_{-\infty}^{\infty} f(x) e^{-i\omega x} dx = F(\omega).$$

Since the autocorrelation is symmetric, it should be no surprise that the function $S(\omega)$ defined below is real.

$$S(\omega) = \int_{-\infty}^{\infty} RC(t) e^{-i\omega t} dt. \tag{3.3.6}$$

Further, using the same logic as above, this can be shown to be an expression of the Weiner-Khinchin theorem. Thus, the Fourier Transform of an autocorrelation function, $S(\omega)$, can be shown to be real and non-negative.

A useful theorem regarding Fourier Transforms concerns the transform of the product of two functions — the convolution function.

$$\int_{-\infty}^{\infty} f(t)g(t)e^{-i\omega t}dt = \int_{-\infty}^{\infty} F(\omega')G(\omega - \omega')d\omega', \qquad (3.3.7)$$

where $G(\omega)$ is the Fourier Transform of $g(t)$. It should be obvious that the role of f and g can be interchanged. One way to look at this result is to consider the situation where F is a "wide" function and G is a "narrow" function. Wide and narrow denote the rms. width of the functions or the half-widths at half-height. In this case, the result of multiplying the two functions and taking the Fourier Transform of the product is to generate a function that looks like $F(\omega)$ smoothed out by $G(\omega)$. The idea here is that G would be non-zero only at values of the integration variable that are close to ω because G is narrow. The result is that the value of the convolution at ω is the value of F at ω weighted by the smoothing function G.

Fourier Transforms have the interesting, if informal, property of transforming functions such that wide functions transform to narrow functions and vice versa. If $1/\Lambda$ is the width of a function in the time domain, its width in the "frequency" domain is approximately Λ.

For example, the function $\exp[-\Lambda|t|]$ (of "width" $1/\Lambda$) transforms to $2\Lambda/(\Lambda^2 + \omega^2)$. In this case, the Fourier Transform has no rms. width, but it obviously has the half width at half height of Λ.

Another example is the Gaussian function, $\exp[-t^2/2\sigma^2]$. It transforms to $A\exp[-\omega^2\sigma^2/2]$. (The constant A is not important in this discussion.) The rms. width of the original signal is σ and the rms.width of the transform is $1/\sigma$.

The reason this property is important to us is that we must deal with the effect of the term $(1 - |\tau|/T)$ in the expressions for the computed spectra from random data, Eq. 3.3.4. The Fourier Transform of this function is

$$2\frac{(1 - \cos[\omega T])}{\omega^2 T}. \qquad (3.3.7)$$

It is a function whose height at the origin is T and the width in the frequency domain is approximately $1/T$. The effect of the term is to smooth the spectrum of the correlation function with a smoothing function whose width is $1/T$. The intrinsic width of the signal of interest whose autocorrelation is being computed does not change with averaging time, whereas the width of the smoothing function gets narrower and narrower. Thus, for T large enough, the effect of this term is negligible. Whatever the width of the interesting features in the signal, $1/T$ is narrower for T sufficiently large.

3.4 Processing Time Series by Computer

At present, the most common method of obtaining and processing data is by means of a digital computer. In the section that follows, I will assume that data is taken at regular intervals, Δt in duration. The data obtained will be denoted $\{f_i\}$, where the data points are numbered in the sequence they are taken — a time series. In particular, I will assume that the data taking device, for instance, an analog to digital converter (ADC), works perfectly, so that no points will be thrown out as being "outliers."

It is easy to approximate the integral used in the computation of the mean, Equation 3.1.1, *viz.*

$$\overline{f}_T \simeq \frac{\Delta t}{T} \sum_{k=1}^{N} f(k) \qquad (3.4.1)$$

where $N = T/\Delta t$. The formula for the variance of the mean remains

$$\sigma_T^2 \simeq \frac{2\sigma^2 \Lambda}{T} .$$

Let $I_{ff} = \Lambda/\Delta t$, the number of lag channels in the integral time scale of f. Then

$$\sigma_T^2 \simeq \frac{2\sigma^2 I_{ff}}{N} . \qquad (3.4.2)$$

The number of independent estimates of the mean is $N/2I_{ff}$.

Example: Suppose $\Lambda \approx 10$ ms. and $\Delta t = 0.1$ ms. Then $I_{ff} = 100$. If further, $N = 1000$, then the number of independent estimates of the mean is $1000/200 = 5$. On the other hand, if $\Delta t = 100$ ms, then $I_{ff} = 1$, and the number of independent estimates of the mean is 500! It is the number of independent estimates of the mean that matters, not the number of data points.

3.5   Estimation of the Autocorrelation

$$\overline{RC}_{ff}(p) = \frac{1}{N} \sum_{i=1}^{N-p} f_i f_{i+p} \tag{3.5.1}$$

$$\langle \overline{RC}(p) \rangle = \frac{1}{N} \sum_{i=1}^{N-p} \langle f_i f_{i+p} \rangle$$

$$= \frac{1}{N} \sum_{i=1}^{N-p} RC(p) = \left(1 - \frac{|p|}{N}\right) RC(p). \tag{3.5.2}$$

Since there is no ambiguity, the Δt is omitted. Again, notice the appearance of the weighting term that comes from finite samples of the data. If $N \gg I_{ff}$, the effect of this term is negligible. The symbol $\overline{RC}(p)$ will be used from now on to denote $RC(p)(1 - |p|/N)$.

The term $(1 - |p|/N)$ will appear so often, it is worth defining it as a special function.

Let

$$W(p, N) \equiv \begin{cases} \left(1 - \dfrac{|p|}{N}\right), & |p| \leq N \\ 0, & |p| > N \end{cases}. \tag{3.5.3}$$

3.5.1   *Error covariance for autocorrelation estimates*

Estimation of the autocorrelation or autocovariance from a data set creates a new time series of random numbers. We will wish to estimate the variance in this set of numbers, i.e., the measurement error. For reasons I will explain later, we will compute the co-correlation of the estimate of the correlation function. There is a possibility that the error in the estimation of the autocorrelation at point p will correlate with the error at point q.

Please note that in this section, we will be trying to estimate quantities from a random process. This is different from trying to estimate some parameter that is partially obscured by noise. In this latter case, you are trying to minimize the effect of a random variable — the noise. The formal definition of the co-correlation is given by

$$\varepsilon^2(p, q) = \langle (\overline{RC}(p) - \langle \overline{RC}(p) \rangle)(\overline{RC}(q) - \langle \overline{RC}(q) \rangle) \rangle$$

$$\varepsilon^2(p,q) = \left\langle \frac{1}{N^2} \sum_{i=1}^{N-p} f_i f_{i+p} \sum_{j=1}^{N-q} f_j f_{j+q} \right\rangle$$

$$- \langle \overline{RC}(p)\rangle\langle \overline{RC}(q)\rangle$$

$$\varepsilon^2(p,q) = \frac{1}{N^2} \sum_{j=1}^{N-p} \sum_{i=1}^{N-q} \langle f_i f_{i+p} f_j f_{j+q}\rangle$$

$$- \langle \overline{RC}(p)\rangle\langle \overline{RC}(q)\rangle. \tag{3.5.4}$$

Once again we have to compute a fourth moment to estimate the measurement error in a second moment. It is not possible to do the general case. The statistics of each different kind of problem must be examined separately.

However, it will be useful to look at the co-correlation of a completely random time series — one that consists of all independent data points. For the sake of simplicity, I assume that the data set has zero mean.

$$\langle RC(p)\rangle = \sigma^2 \delta_{0p}, \tag{3.5.5}$$

since the covariance of all the cross terms is zero. It should be obvious that for this data set, $I_{ff} = 1$.

We use 3.5.4 to calculate the co-correlation.

$$\langle f_i f_{i+p} f_j f_{j+q}\rangle = \langle f_i\rangle\langle f_{i+p}\rangle\langle f_j\rangle\langle f_{j+q}\rangle = 0,$$

unless two or more of the indices are equal. This term is still zero, even if some of the indices are equal. Consider that case where $i = j$, but $p \neq q$. I also assume p and q are not zero.

$$\langle f_i f_{i+p} f_i f_{i+q}\rangle = \langle f_i^2\rangle\langle f_{i+p}\rangle\langle f_{i+q}\rangle = 0.$$

It can be seen that all terms where $p \neq q$ are zero.

Because of the independence of different points in the time series, the terms that remain in the sum in 3.5.4 are the terms where $p = q$ and of the form $i = j, p$ any value except zero; $i \neq j, p = 0$; and $i = j, p = 0$.

For $i = j$, arbitrary p, we get

$$\langle f_i f_i f_{i+p} f_{i+p}\rangle = \langle f_i f_i\rangle\langle f_{i+p} f_{i+p}\rangle = \langle \sigma^2\rangle^2.$$

There are N terms of that form, for each p.

For the second case, $i \neq j, p = 0$,

$$\langle f_i f_i f_j f_j\rangle = \langle f_i f_i\rangle\langle f_j f_j\rangle = \delta_{0p}\langle \sigma^2\rangle^2.$$

There are $N(N-1)$ terms of this form.

Finally, there are N terms of the form $i = j$, $p = 0$.

$$\langle f_i f_i f_i f_i \rangle = \langle f^4 \rangle .$$

Adding all this up, we get

$$\varepsilon^2(p) = \frac{1}{N}\langle f_i^4 \rangle \delta_{0p} + (1 - \frac{1}{N})\delta_{0p}(\sigma^2)^2 + (1 - \delta_{0p})\frac{(\sigma^2)^2}{N} - \delta_{0p}(\sigma^2)^2$$

$$(3.5.6)$$

$$\varepsilon^2(p) = \frac{1}{N}[(\langle f^4 \rangle - (\sigma^2)^2)\delta_{0p} + (1 - \delta_{0p})(\sigma^2)^2] .$$

If $\{f_i\}$ has Gaussian statistics, then $\langle f^4 \rangle = 3(\sigma^2)^2$, and the measurement variance can be written

$$\varepsilon^2(p) = \frac{2}{N}\langle \sigma^2 \rangle^2, \ p = 0 ,$$

$$(3.5.7)$$

$$\varepsilon^2(p) = \frac{1}{N}\langle \sigma^2 \rangle^2, \ p \neq 0 .$$

Notice that there is a nonzero measurement variance, even for those points where the expected correlogram is zero.

You should always be careful to make a distinction between the statistics of the original signal and the statistics of the measured correlogram. As I have been stressing, the statistics of the original signal may not be Gaussian. However, since the estimated correlogram is made up of the sums of many products of random variables, the points in the measured autocorrelation will usually have Gaussian statistics because of the central limit theorem.

Having made the obligatory warnings, allow me to look at a case of a non-trivial time series that does have Gaussian statistics. We will not assume a lack of correlation between points in the autocorrelation. In this case we can decompose the fourth moment into a product of second moments, *viz.*

$$\langle f_i f_j f_l f_m \rangle = \langle f_i f_j \rangle\langle f_l f_m \rangle + \langle f_i f_l \rangle\langle f_j f_m \rangle + \langle f_i f_m \rangle\langle f_j f_l \rangle - 2\langle f_i \rangle^4 .$$

Then 3.5.4 can be written

$$\varepsilon^2(p, q) = \frac{1}{N^2} \sum_{j=1}^{N-p} \sum_{i=1}^{N-q} [RC(i - j)RC(i - j + p - q)$$

$$+ RC(i - j - p)RC(i - j + q)] - 2\langle f \rangle^4 .$$

We can subtract the $\langle f \rangle^4$ term and rewrite the expression as

$$\varepsilon^2(p, q) = \frac{\sigma^4}{N^2} \sum_{j=1}^{N-p} \sum_{i=1}^{N-q} [R(i - j)R(i - j + p - q)$$

$$+ R(i - j - p)R(i - j + q) + \frac{\langle f \rangle^2}{\sigma^2}(R(i - j)$$

$$+ R(i - j + p - q) + R(i - j - p) + R(i - j + q)].$$

The terms in the third line arise because of the uncertainty in the estimation of the mean, $\langle f \rangle$. Most authors assume a zero mean when doing this calculation and these terms do not appear. But, in some experiments you cannot have the luxury of working with a zero mean signal and I decided, here, to leave those terms in.

Now, we change summation variables and get

$$\varepsilon^2(p, q) = \frac{\sigma^4}{N} \sum_{k=-N}^{N} W(p, N)W(q, N)[R(k)R(k + p - q)$$

$$+ R(k - p)R(k + q) + \frac{\langle f \rangle^2}{\sigma^2}(R(k)$$

$$+ R(+p - q) + R(k - p) + R(k + q)], \quad p > q. \quad (3.5.8)$$

Typically, a covariance function is close to one for small values of its argument (compared to I_{ff}) and near zero for large values of its argument. By taking account of this behavior, we can get a general idea of how the covariance of the measurement error behaves without knowing an explicit form for the expected autocorrelation or autocovariance.

At $p \approx q \approx 0$, the sums above can be approximated by

$$\varepsilon^2(p, q) \approx \frac{\sigma^4 I_{ff}}{N}\left[4R(p - q) + 8\frac{\langle f \rangle^2}{\sigma^2}\right]. \quad (3.5.9)$$

For p and q large, but $p \approx q$, the sums are approximately given by

$$\varepsilon^2(p,q) \approx \frac{\sigma^4 I_{ff}}{N}\left[2R(p-q) + 8\frac{\langle f \rangle^2}{\sigma^2}\right]. \qquad (3.5.10)$$

Several essential things can be seen from these calculations:

(1) The measurement error correlates over the measured correlogram; i.e., if the ith measured point is high, the $i+1$ point is probably also high. The correlation length of the errors is approximately I_{ff}. This is important, because if you are interested in estimating some kind of mean quantity from a measured correlogram, you will have fewer independent data points than the number of points in the measured correlogram.

(2) The measurement error is larger near the origin than near the tail, but never goes to zero, even if the expected correlation function does.

(3) A non-zero baseline can have a large effect on the measurement variance. Subtracting an estimated mean from the signal before correlating will not help. Recall, we have computed the variance from the expected signal, not the variance from the average signal.

For some forms of the autocorrelation function, it is possible to estimate the sums in 3.4.8 by integrals that can be performed. For instance, let $\langle R(m) \rangle = A \exp[-\gamma m]$ be the autocorrelation for a zero mean signal. Then, the covariance of the measurement error can be shown to be given by

$$\varepsilon^2(p,q) \simeq \qquad (3.5.11)$$

$$\frac{A^2(\exp(-\gamma|p+q|)(1+\gamma|p+q|) + \exp(-\gamma|p-q|)(1+\gamma|p-q|))}{\gamma N}.$$

Exercise:
Using the uniform generator modified to have zero mean and a variance of one, generate 1000 points. Compute 20 lag times of the autocorrelation function, starting at a lag of zero. The expected correlogram is given by 3.4.5. Examine the variance of the measurement by computing the square of the difference between the measured correlogram and the expected correlogram as a function of the lag time.

Using the Gaussian generator, generate 1000 points. Compute 20 lag times of the autocorrelation function, starting at a lag of zero. Again, the expected correlogram is given by 3.5.5. Examine the variance of the measurement by computing the square of the difference between the measured correlogram and the expected correlogram as a function of the lag time. Is there the expected difference between this measured variance and that for the uniform generator? If you cannot tell, how many points do you need to correlate before you can tell?

3.5.2   *Random, correlated signal generator*

Until now, the computer noise generators used have been carefully designed to give uncorrelated outputs. In order to examine the theory talked about in this and in subsequent sections, we need a correlated time series generator. There are many such generators. The simplest one I know is the following:

(1) Generate N random numbers using the uniform or the Gaussian generator. Denote the points generated, $\{x_n\}$.
(2) Pick a number between 0 and 1, call it α.
(3) Make a new time series, $\{y_n\}$, from the old, by

$$y_n = \alpha x_n + (1 - \alpha)y_{n-1}.\qquad (3.5.12)$$

This time series has several interesting features:

(1) Looked upon as a filter for the original series, it has an impulse response of

$$y_n = \alpha(1 - \alpha)^n x_0.$$

(2)

$$\langle y_n \rangle = \langle x_n \rangle.$$

(3)

$$\sigma_y^2 = \frac{\alpha}{2 - \alpha}\sigma_x^2.$$

(4)

$$R_{yy}(k) = (1 - \alpha)^{|k|} = e^{|k| \ln(1-\alpha)}.$$

(5)

$$RC_{yy}(k) = \frac{\alpha \sigma_x^2}{2 - \alpha}e^{|k| \ln(1-\alpha)} + \langle x \rangle^2.\qquad (3.5.13)$$

Exercise:
Using the uniform generator with a zero mean and a variance of 1, generate 1000 points for a new series using the above algorithm, having $\alpha = 0.2$.

Examine the pdf of the new data set. Is it Gaussian? Can you tell? Now generate 12,800 numbers of the series, 128 at a time. Compute the autocorrelation of the data set, 128 points at a time. Use the algorithm 3.4.1, with $N = 128$. Then add up 100 such computations and divide the result by 100. The effective number of points averaged will be 12,800. Can you see the effect of the term $(1 - \frac{k}{N})$?

(1) Create the measurement variance series

$$\left(\overline{RC}(k) - \langle \overline{RC}(k) \rangle \right)^2 .$$

(2) Compare these results to formula 3.5.9 and 3.5.10. Note, in particular, that the errors in nearby channels correlate.

3.6   Estimation of the Power Spectrum of Time Series Data

The definition of the digital complex Fourier Transform (DFT) of the digital data set $\{x_n\}$ is

$$s_N(m) = \sum_{k=0}^{N-1} x_k \exp\left(-i\frac{2\pi km}{N} \right). \qquad (3.6.1)$$

The average **power spectrum** is defined by

$$S_N(m) = \frac{1}{N} |s_N(m)|^2 . \qquad (3.6.2)$$

$$S_N(m) = \frac{1}{N} \sum_{k=0}^{N-1} \sum_{l=0}^{N-1} x(k)x(l) \exp\left(-i\frac{2\pi km}{N} \right) \exp\left(+i\frac{2\pi lm}{N} \right),$$

$$S_N(m) = \frac{1}{N} \sum_{k=0}^{N-1} \sum_{l=0}^{N-1} x(k)x(k + (l - k)) \exp\left(-i\frac{2\pi m(k - l)}{N} \right).$$

Now changing the summation variables and doing the k sum first, we get

$$S_N(m) = \sum_{p=-N-1}^{N-1} \overline{RC}(p) \exp\left(-i\frac{2\pi mp}{N}\right). \tag{3.6.3}$$

Computing the average power spectrum is essentially the same as computing the DFT of the estimated autocorrelation function. This is the digital version of the Weiner–Khinchin theorem.

$$\langle S_N(m)\rangle = \sum_{p=-N-1}^{N-1} W(p, N)\langle RC(p)\rangle \exp\left(-i\frac{2\pi mp}{N}\right). \tag{3.6.4}$$

However, be careful because there are some problems with the DFT. The DFT is usually defined as a one-sided sum, i.e. from 0 to $N - 1$. (More to the point, most canned DFT routines go from 0 to $N - 1$.) The expressions above are two-sided sums, from $-N - 1$ to $N - 1$. Since the autocorrelation function is, by definition, symmetrical, you can get the equivalent result from a typical one-sided DFT by

$$S_N(m) = 2\mathrm{Real}\left[\sum_{p=0}^{N-1} \overline{RC}(p) \exp\left(-i\frac{2\pi mp}{N}\right)\right] - \overline{RC}(0). \tag{3.6.5}$$

The astute reader has probably asked why we do the full Digital Fourier Transform and then take the real part of the result instead of just doing the equivalent Cosine Transform. This will work. The reason for doing the Digital Fourier Transform is the existence of so-called Fast Fourier Transform routines. The concept was first put forth by Cooley and Tukey [Mathematics of Computation, vol. 19, no. 90, pp 297–301 (1965)] The result is that it is possible to do the complex Fourier Transform using $N \ln_2 N$ multiplications instead of the roughly N^2 multiplications necessary for a traditional Cosine Transform. This drastically cuts down on the computational time for large data sets.

3.6.1 *Generic properties of the digital Fourier transform*

Even more important is the fact that the DFT is a digitized copy of the Fourier Transform of the original function only under very specific circumstances. This inconvenient fact has too often been ignored.

I will first demonstrate using the classical problem, namely the transform of $\exp[-\lambda|t|]$. It is well known that the Fourier Transform of this function is

$$\frac{2\lambda}{\lambda^2 + \omega^2} \, .$$

The digitized sample of this function would be obtained by substituting $\omega = 2\pi m/N$, *viz.*

$$\text{FT}(e^{-\lambda|n|}) = \frac{2\lambda}{\lambda^2 + (2\pi m/N)^2} \, .$$

On the other hand, we can get a closed form for the digital transform of the exponential,

$$\text{DFT}(e^{-\lambda|n|}) = \frac{1 - 2e^{-N\lambda} - e^{-2\lambda} + 2e^{-(N+1)\lambda}\cos(2\pi m/N)}{1 + e^{-2\lambda} - 2e^{-\lambda}\cos(2\pi m/N)} \, .$$

This is hardly the same function! Among other things, the DFT is a periodic function with a period, N. The problem can be looked at as coming from the fact that the DFT process is really a Finite Fourier Series computation. It can be shown that the DFT is related to the FT by

$$\text{DFT}(m) = \sum_{k=-\infty}^{\infty} \text{FT}(m - kN) \, . \tag{3.6.6}$$

This phenomenon is called aliasing. Therefore unless a signal has no energy at frequencies over $1/(2\Delta t)$, where Δt is the sampling interval, the DFT will be distorted by the energy leaking into any one frequency range from the other copies of the FT. The normal method of dealing with this problem is to filter the input signal to cut off all frequency content above $1/(2\Delta t)$.

The required cut-off frequency, $1/(2\Delta t)$, is popularly known as the Nykvist frequency.

Other interesting properties of the DFT:

$$\sum_{n=0}^{N-1} f(n)g(n)\exp\left[-i\frac{2\pi nm}{N}\right] = \sum_{k=0}^{N-1} F(k)G(m - k) \, , \tag{3.6.7}$$

where $F(m)$ is the DFT of $f(n)$.

$$\frac{1}{N} \sum_{m=0}^{N-1} F(m) e^{i\frac{2\pi nm}{N}} = f(n).$$ (3.6.8)

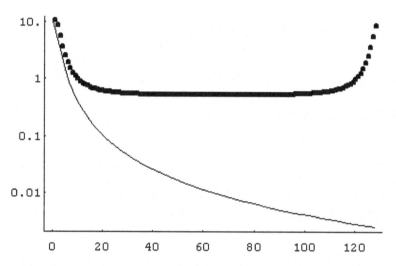

Fig. 3.3 Fourier transform (light line) and DFT (heavy line) of $\exp(-0.1n)$ for 128 points.

Exercises:

1. Prove that $S_N(m) = S_N(-m)$. Is this true for the DFT of any function? Prove that

$$S(0) = \sum_{n=0}^{N-1} f(n).$$

3.6.2 *The variance of spectrum estimates*

We can now examine the covariance of the spectrum estimates.

$$\varepsilon_S^2(m, n) = \langle (S^2(m) - \langle S^2(m) \rangle)(S^2(n) - \langle S^2(n) \rangle) \rangle$$

$$\varepsilon_S^2(m,n) = \sum_{p,q=-N-1}^{N-1} \sum e^{-i\frac{2\pi mp}{N}} e^{i\frac{2\pi nq}{N}} (\langle \overline{RC}(p)\overline{RC}(q)\rangle$$

$$- \langle \overline{RC}(p)\rangle\langle \overline{RC}(q)\rangle) . \tag{3.6.9}$$

Once again, we need to compute the fourth moment of a time series. And once again, there is no simple closed form for all such series. However, it is again useful to examine a signal with Gaussian statistics. In this case, we can substitute the expression we have already computed for the fourth moment, 3.5.8, into the equation above, and perform the indicated operations. We get

$$\varepsilon_S^2(m,n) \simeq S^2(m)\delta_{nm} . \tag{3.6.10}$$

Huh!? Yes, the variance of the spectrum estimate computed as in 3.5.4, is the same as the square of the spectrum. It may not be obvious yet, but the problem is that the Fourier Transform was taken of the entire N data points.

There is another way to compute the spectrum. Take sub-records of length of $N_s < N$. Compute the spectrum of this record using 3.6.2. Do that for N/N_s times and average the results. It can be shown that the variance of this estimate is

$$\varepsilon_S^2(m) \simeq \frac{N_s}{N}(\tilde{S}^2(m)\delta_{nm}) \tag{3.6.11}$$

where $\tilde{S}(m)$ is the DFT of $RC(n)(1 - \frac{|n|}{N_s})$. The proof is simply to note that each DFT is an independent sample of the spectrum. Averaging N/N_s samples decreases the variance of the estimate by the number of samples.

What has happened here is that the spectrum information has been obtained from a subset of the possible frequencies. This is equivalent to smoothing of the data after computation of the Fourier Transform, *viz.*

$$\tilde{S}(m) = \sum_{-(N-1)}^{N-1} W(n,N_s)RC(n)e^{-i\frac{2\pi nm}{N}} .$$

By 3.6.7, the Convolution Theorem, this expression can be rewritten

$$\tilde{S}(m) = \sum_{k=-(N-1)}^{N-1} \hat{W}(m-k, N_s)S(k), \qquad (3.6.12)$$

where $\hat{W}(\,)$ is the Fourier Transform of $W(\,)$. The function $\hat{W}$, smooths the Transform $S(k)$.

A careful reading of the above section should convince you that the smoothing function is necessary to getting good estimates of the spectrum of random signals. In electrical engineering this smoothing process is called windowing. The price you pay for windowing is a decrease in the frequency resolution of the resulting spectrum. The bandwidth of the smoothing filter is $\simeq 1/(N_s\Delta t)$, so the finest discernible feature will have to be wider in bandwidth than that. There is no point in having a finer frequency scale than the bandwidth of the smoothing filter, so most modify the calculation to take place only over the range where $W(k, N_s)$ is not zero and with a frequency step of the bandwidth of the filter, *viz.*,

$$\tilde{S}(m) = \sum_{n=-(N_s-1)}^{N_s-1} W(n, N_s)RC(n)e^{-i\frac{2\pi nm}{N_s}}. \qquad (3.6.13)$$

An additional bonus is that the number of computations is decreased. A FFT routine takes $N \ln_2 N$ multiplications and additions to compute the DFT. Computation of the correlogram indicated above takes $N_s N$ multiplications and additions and then computation of the spectrum takes $N_s \ln_2 N_s$ more computations. The latter method usually involves fewer computations than the FFT of the full data set, followed by smoothing.

Exercises:
Using the zero mean uniform generator, and then the correlated noise generator, with an α of 0.2, generate time series, 128 points at a time.

(1) Compute the complex spectrum of the set of 128 points.
(2) Compute the power spectrum from the complex spectrum.
(3) Compute the 128 lag channel autocorrelation for the set of points.
(4) Repeat the above 100 times and average the results for each of the spectra and the autocorrelation. Compute the power spectrum from the autocorrelation function and from the average complex spectrum. Be careful to use Equation 2.5.5. Also, watch the scaling on the average spectra. If you are not careful, you will divide by 100 twice. Which

has the smallest variance? Is there any difference at all between the average power spectrum computed by "b" and by transforming the autocorrelation? Method "a" above is equivalent to computing the power spectrum for the entire data set.

3.7   Batch Mode Autocorrelation

There is an alternate algorithm for use in estimating the autocorrelation and autocovariance of time series called Batch Mode Autocorrelation. The algorithm is

$$\overline{RCB}(m) = \frac{1}{N_T} \sum_{k=0}^{N_T} f(kN_s)f(kN_s + m)\,, \qquad (3.7.1)$$

where N_s is usually picked to be larger than $2\,I_{ff}$. You take a data record N_s points long. Multiply the first point times each of the other points and add the results to the appropriate lag channel. Then discard the data set and take another set. Repeat the procedure N_T times.

This differs from the previous procedure in that each batch of data has only one set of multiplications performed in it. Previously, we multiplied the first point times all the points, the second point times all but the first point, etc. Before there were

$$(N_T^2 + N_T)/2$$

multiplications per batch of data. In the new method, there are only N_T multiplications per batch of data.

$$\langle \overline{RCB}(m) \rangle = \frac{1}{N_T} \sum_{k=0}^{N_T} \langle f(kN_s)f(kN_s + m) \rangle\,,$$

$$\langle \overline{RCB}(m) \rangle = \frac{1}{N_T} \sum_{k=0}^{N_T} RC(m) = RC(m)\,. \qquad (3.7.2)$$

The expected value is the true autocorrelation function, except that unlike before, the window function is missing. Recall that the window arose because of the finite length of the data record allowing only $N - m$ multiplications and additions for the mth lag channel. Here, there is no such problem. But, what about the variance of the estimate of the correlogram?

$$\varepsilon^2(p,q) = \left\langle \frac{1}{N_T^2} \sum_{j=0}^{N_T-1} f(jN_s)f(jN_s+q) \right\rangle^2 - \langle \overline{RCB}(p) \rangle \langle \overline{RCB}(q) \rangle$$

$$\varepsilon^2(p,q) = \frac{1}{N_T^2} \sum_{l=0}^{N_T-1} \sum_{j=0}^{N_T-1} \langle f(lN_s)f(lN_s+p)f(jN_s)f(jN_s+q) \rangle$$

$$- \langle \overline{RCB}(p) \rangle \langle \overline{RCB}(q) \rangle .$$

Since N_s is picked to be larger than $2I_{ff}$, when $j \neq i$, the data sets are statistically independent, so the computation breaks into two parts, *viz.*

$$\langle f(lN_s)f(lN_s+p)f(hN_s)f(jN_s+q) \rangle$$
$$= \langle f(lN_s)f(lN_s+p) \rangle \langle f(jN_s)f(jN_s+q) \rangle, \quad l \neq j .$$

There are $N_T(N_T+1)$ terms of this type. And,

$$\langle flN_s)f(lN_s+p)f(jN_s)f(JN_s+q) \rangle$$
$$= \langle f^2(lN_s)f(lN_s+p)f(lN_s+q) \rangle, \quad l = j .$$

There are N_T terms of this type.
But,

$$\langle f(lN_s)f(lN_s+p) \rangle \langle f(jN_s)f(jN_s+q) \rangle = \langle \overline{RCB}(p) \rangle \langle \overline{RCB}(q) \rangle .$$

The expression for the covariance simplifies to

$$\varepsilon^2(p,q) = \frac{1}{N_T} \langle (f(lN_s)^2 f(lN_s+p)f(lN_s+q) \rangle$$
$$- \langle \overline{RCB}(p) \rangle \langle \overline{RCB}(q) \rangle) .$$

The results to here are independent of the statistics of f. As usual, in order to evaluate this expression we need a formula for the fourth moment. If we assume Gaussian statistics, we can get a feel for the behavior of the process. Making this assumption and using the formula for the fourth moment, we get

$$\varepsilon^2(p,q) = \frac{1}{N_T} (\langle RCB(0) \rangle \langle RCB(p,q) \rangle + \langle RCB(p) \rangle \langle RCB(q) \rangle) .$$

$$(3.7.3)$$

Recall that N_s is picked to be larger than I_{ff}, the process correlation time. Let us assume for the moment that it was picked to be on the order of $2I_{ff}$. The total record length is N and $N_T = N/N_s \approx N/2I_{ff}$. If this is substituted into the formula for the variance, you get a result that is very similar to what you got when using the full autocorrelation computation.

$$\varepsilon^2(p,q) = \frac{2I_{ff}}{N}\sigma^4(R(0)R(p-q) + R(p)R(q)), \qquad (3.7.4)$$

where R is again the autocovariance. Compare this result to 3.4.9 and 3.4.10. The variance of the estimated correlogram is about the same as that obtained doing all the multiplications. Many commercial correlation computers take advantage of this fact and actually implement the batch mode for computation of correlograms. Since the batch method requires many fewer multiplications, the processor can be built with a much smaller bandwidth than would otherwise be necessary.

In retrospect, it should not be surprising that the batch method is so efficient. We have seen all along that measurement variances decrease as the number of independent estimates increase. When you use the full auto-correlation routine, the product of the first and second points is not usually statistically independent of the product of the second and third points and so on. Using the batch mode, each multiplication is statistically independent.

Suggested Reading:

"RandomData: Analysis and Measurement Procedures", J. S. Bendat and A. G. Piersol, 2nd edition, Wiley Interscience, 1986.

Chapter 4

Parameter Estimation

4.1 Motivation

Much of the time an engineer or scientist makes measurements of some quantity because, in fact, the value of another quantity is desired. For instance, you may make measurements of temperatures and flow rates of liquids in a tube because you want to know the heat transfer coefficient in the system. You may measure voltages in a circuit because you want to know the gain or linearity of some device. What links the measurements you make to the desired quantities is a physical model. In the cases hinted at above, the measurements are put into the physical model and the desired quantity solved for.

What makes life interesting is either errors in the measurements or noise in the system. Consider a simple example where you take some data that is supposed to be a straight line. The slope of that line is the desired parameter. If you take just two measurements, you have no problem. You have two measurements and two unknowns. Surely you can find some computer package that can handle that. Suppose, however, you take three points and find that the three do not lie exactly on a straight line. Then, what?

This chapter is devoted to attempting to deal rationally with problems of the latter kind. Because of noise or uncertainty in the measurement procedure, the problem is ill posed. Something else needs to be brought to the problem in order to make the problem well posed and soluble.

Suppose you have a model of what a data set should look like: a model function

$$f(n, \boldsymbol{\alpha}).$$

The independent variable is denoted n and $\boldsymbol{\alpha}$ represents the set of parameters of the model function. For instance, in a radioactive decay experiment, $f(\ )$ would be given by

$$f(n, \boldsymbol{\alpha}) = \alpha_0 \exp(-\alpha_1 n \Delta t), \qquad (4.1.1)$$

where α_0 is the initial decay rate, α_1 is related to the half-life of the material and Δt is the time interval for each measurement count. The parameters to estimate are α_0 and α_1.

You could try to obtain the parameters α by simply varying α_0 and α_1 until you match the data exactly. Of course, you can never do it that easily for the simple reason that the data always has "noise" in it. In practice, you vary the coefficients α until you have obtained a "best" fit in some sense. Least Squares is a popular methodology for doing this, although it can be shown that in this example that would not always be an optimal procedure.

Before going into the formal structure of parameter estimation schemes, I will give an example of a typical problem and an example of how *not* to solve the problem. Consider a simple heat transfer experiment wherein a cylinder full of a liquid is allowed to cool from an initial temperature T_0 into a room at T_∞. The temperature variation with time is expected to follow the formula

$$T - T_\infty = (T_0 - T_\infty) \exp\left[-\frac{4ht}{\rho C_p D}\right].$$

The parameters ρ, D, T_∞, and C_p are assumed to be known. The point of the experiment is to measure h, the heat transfer coefficient. In the terms used in this section, the model function is written

$$\alpha_0 \exp(-\alpha_1 t). \qquad (4.1.2)$$

Engineers are usually taught that the way to do this problem is to take the logarithm of the data and plot it vs. t. The resulting plot should be a straight line whose parameters are then determined by using a least squares procedure. That procedure made some sense in the days before computers and handheld calculators. I will show here that this procedure is incorrect and is guaranteed to give the wrong answers! In the following, I assume that the data were taken by sampling at regular time intervals, so that the time will be identified by the variable $n, n = 0, 1, 2, 3 \ldots$.

In any real experiment, there will be "noise" on the data. In the experiment described here, the chief source of errors is the imprecision in reading the thermometer. Such errors tend to be approximately Gaussian with a constant variance; *i.e.*,

$$d(n) = \alpha_0 \exp(-\alpha_1 n) + \varepsilon_n,$$

where ϵ_n is the error at the nth point. Saying that the statistics are Gaussian is to say the pdf for ϵ_n is given by

$$p(\varepsilon_n) = \frac{1}{\sqrt{2\pi\sigma^2}} \exp\left[-\frac{\varepsilon_n^2}{2\sigma^2}\right]. \tag{4.1.3}$$

In this example, the variance of the error is assumed to be independent of the time n. The pdf for the log of the data was derived in Section 1.9; however, I will use an approximate model here to demonstrate the problem with using the logarithm.

Take the logarithm of the data:

$$\log(f(n)) = \log(\alpha_0 \exp(-\alpha_1 n) + \varepsilon_n)$$

$$= \log\left((\alpha_0 \exp(-\alpha_1 n))\left(1 + \frac{\varepsilon_n}{\alpha_0} \exp(\alpha_1 n)\right)\right)$$

$$= \log(\alpha_0) - \alpha_1 n + \log\left(1 + \frac{\varepsilon_n}{\alpha_0} \exp(\alpha_1 n)\right). \tag{4.1.4}$$

The noise term for the logarithm of the data, $\log[1 + \varepsilon_n/\alpha_0) \exp(\alpha_1 n)]$, is clearly a function of time, n, and does not have a constant variance. Note that for large values of n, the term $(\varepsilon_n/\alpha_0) \exp(\alpha_1 n)$ can approach ± 1. For a $+1$ variation, nothing extraordinary happens, but for a fluctuation of the size -1, you have an error term of the order $\log(0)(= -\infty)$! In other words, negative fluctuations are heavily emphasized over positive fluctuations. Further, the fluctuations can make $f(n)$ negative, whereupon you can not take the logarithm. You are forced to eliminate all points with negative points or truncate the data at the first occurrence of a negative point. I object to either option, since experience shows that taking data is very expensive and to throw out data points is even more costly because it will bias the results of the experiment. Table 4.1 contains the results of a computer simulated heat transfer experiment. It was created by first calculating the exact data for a model function

$$f(n) = 100 \exp(-0.3n).$$

Then, zero mean Gaussian noise of expected variance 16 was added to the data. Figure 4.1 shows a plot of the log of the data truncated at the first negative point.

Table 4.1

102.6	72.76	54.54	41.42	24.54	21.42	15.25	12.2	16.44	3.955
11.73	3.131	1.711	0.2614	−7.909	−3.411	2.848	6.415	0.5165	−0.4107

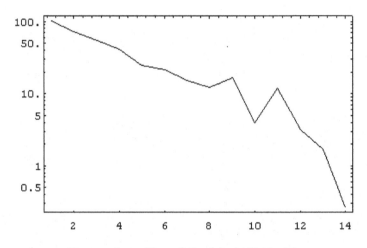

Fig. 4.1   Logarithm of the data in Table 4.1.

Using a least squares program on the logarithm of the data, I obtained values for the parameters

$$\alpha_0 = 127, \quad \alpha_1 = 0.3595.$$

Not bad, but it suggests the bias I mentioned above. The values for α_0 and α_1 are on the high side.

There are really two problems to deal with here. It is still possible to use a least squares procedure, but it will be a nonlinear one since it will have to deal with the nonlinear coefficient α_1. Second, it is not obvious that least squares is the correct procedure to use. At this point, all we know is that least squares is popular and the software to do it is readily available. In what follows, a plausible framework for arriving at the correct procedure to handle problems of this sort will be derived. Later, procedures for dealing with the nonlinear equations derived will be demonstrated.

4.2   Maximum Likelihood Estimation

Because of the noise or uncertainty involved in our measurement procedure, we cannot extract the desired parameter(s) *with certainty*. The answer can only be obtained in some probabilistic sense; i.e., *the answer obtained is probably the best we can get, knowing what we know*. In such an instance it is also very valuable, some say essential, to have some idea of how far off you probably are in your estimate, given what you know.

A method to attack both problems would be to try to construct *the probability of a parameter, given the data that you have collected*. A conditional probability arises because the "best" answer will clearly depend on what data you have. In that case, you would have a well posed problem to solve and having the probability function, you could estimate the measurement variances. First consider the joint probability of the data and the parameters,

$$p(\boldsymbol{\alpha}, \{d_i\}),$$

where $\boldsymbol{\alpha}$ is a vector of parameters to be estimated, and $\{d_i\}$ represents a set of data points. From Bayes theorem, (See Equation 1.8.1) this probability is related to conditional and other probabilities by

$$p(\boldsymbol{\alpha}, \{d_i\}) = p(\{d_i\}|\boldsymbol{\alpha})p(\boldsymbol{\alpha}) = p(\boldsymbol{\alpha}\,|\{d_i\})p(\{d_i\})$$

$$p(\boldsymbol{\alpha}|\{d_i\}) = \frac{p(\{d_i\}|\boldsymbol{\alpha})p(\boldsymbol{\alpha})}{p(\{d_i\})}\,. \tag{4.2.1}$$

I have expressed the conditional probability of the parameters, given the data, in terms of the conditional probability of the data, given the parameters and the probabilities of the data and the parameters. It turns out that often $p(\{d_i\}|\boldsymbol{\alpha})$ is relatively easy to compute. Further, $p(\{d_i\})$ is not a function of the parameters, so it will not be involved in any procedure to maximize the probability of the parameters. Finally, the probability of the parameters, $p(\boldsymbol{\alpha})$, needs to be dealt with. In many problems, this probability is not known *a priori* (ahead of time). At best, a range of possible values for the parameters is known. In this case, the probability of the parameters is not an explicit part of the optimization and the term is left out of the equations. When this is the procedure, it is called *Maximum Likelihood Estimation*, MLE. In the case where $p(\alpha)$ is known, the procedure is called *Maximum A Posteriori Estimation*, MAP .

The secret to finding the probability of a data set is to have a model for the measured fluctuations in the data; *i.e.*, the noise. In the heat transfer experiment described above the errors ϵ had a Gaussian pdf. In that example, $p(d_n|\boldsymbol{\alpha})$ is given by

$$p(d_n|\boldsymbol{\alpha}) = p(\varepsilon) = \frac{1}{\sqrt{2\pi\sigma^2}} \exp\left(-\frac{(d_n - f(n,\boldsymbol{\alpha}))^2}{2\sigma^2} \right), \qquad (4.2.2)$$

where

$$f(n,\boldsymbol{\alpha}) = \alpha_0 \exp(-\alpha_1 n).$$

Commonly, the data at each measurement point are statistically independent of the data at every other interval, thus the errors are independent random variables. Then using the formula derived in chapter 1 for the joint probability of multiple independent random variables, we get the probability of the data set given α as

$$p(\{d_n\}\,|\boldsymbol{\alpha}) = p(d_1|\boldsymbol{\alpha})p(d_2|\boldsymbol{\alpha})p(d_3|\boldsymbol{\alpha})\cdots = \prod_n p(d_n|\boldsymbol{\alpha}). \qquad (4.2.3)$$

Since the logarithm of a function is a maximum when the function is a maximum, most users prefer to use the logarithm of the probability. The logarithm of the probability of the data set is given by

$$\log p(\{d_n\}|\alpha) = \sum_n \log p(d_n|\alpha). \qquad (4.2.4)$$

Finding the set of parameters α that will maximize the probability of the data set is done by solving the set of equations

$$\frac{\partial \log p(\{d_n\}|\alpha)}{\partial \alpha_i} = 0, \qquad i = 1, 2, 3, \ldots, \qquad (4.2.5)$$

subject to the usual constraints that the second derivatives be negative. These are called *The Maximum Likelihood Equations* . Since we have the data set $\{d_n\}$ and the model function $f(n,\boldsymbol{\alpha})$, and an explicit function for $p(\{d_n\}|\boldsymbol{\alpha})$, the problem is now completely specified.

4.2.1   *Gaussian processes*

Let us see what the explicit equations look like for the heat transfer experiment.

$$\log p(d_n|\boldsymbol{\alpha}) = -\frac{(d_n - f(n,\boldsymbol{\alpha}))^2}{2\sigma^2} - 0.5\log 2\pi\sigma^2$$

$$\log p(\{d_n\}|\boldsymbol{\alpha}) = -\sum_n \frac{(d_n - f(n, \boldsymbol{\alpha}))^2}{2\sigma^2} - 0.5 \sum_n \log 2\pi\sigma^2. \qquad (4.2.6)$$

Since in this example σ is not a function of $\boldsymbol{\alpha}$, this problem is solved by maximizing the first sum in the expression above, or minimizing *minus* the expression. This is exactly a least squares procedure. We have shown that for **a measurement process with constant Gaussian errors, the MLE method is identical to a least squares procedure.**

If the variance of the measurements, σ^2, was a function of α and/or n, the method would be a *weighted least squares* procedure. In that case, the terms involving derivatives of σ^2 would have been kept.

$$\log p(\{d_n\}|\boldsymbol{\alpha}) = -\sum_n \frac{(d_n - f(n, \boldsymbol{\alpha}))^2}{2\sigma_n^2} - 0.5 \sum_n \log 2\pi\sigma_n^2. \qquad (4.2.7)$$

In terms of the heat transfer experiment discussed earlier, the MLE equations are

$$\frac{\partial \log p(\{d_n\}|\boldsymbol{\alpha})}{\partial \alpha_0} = 2 \sum_n \frac{\partial f(n, \boldsymbol{\alpha})}{\partial \alpha_0} \left(\frac{(d_n - f(n, \boldsymbol{\alpha}))}{2\sigma^2} \right)$$

$$= \frac{1}{\sigma^2} \sum_n e^{-\alpha_1 n}(d_n - \alpha_0 e^{-\alpha_1 n}) = 0,$$

$$\frac{\partial \log p(\{d_n\}|\boldsymbol{\alpha})}{\partial \alpha_1} = 2 \sum_n \frac{\partial f(n, \boldsymbol{\alpha})}{\partial \alpha_1} \left(\frac{(d_n - f(n, \boldsymbol{\alpha}))}{2\sigma^2} \right)$$

$$= -\frac{\alpha_0}{\sigma^2} \sum_n n e^{-\alpha_1 n}(d_n - \alpha_0 e^{-\alpha_1 n}) = 0.$$

Using these expressions, I computed the estimates of the parameters for the data in Table 4.1 to be

$$\alpha_0 = 101.3 \quad \text{and} \quad \alpha_1 = 0.30987.$$

Notice that these estimates are much closer to the "correct" values than the estimates obtained by taking the logarithm and doing a linear least squares fit on the same data set. I put correct in quotes to remind you that you do not usually know the correct answer.

4.2.2   *Poisson processes*

However, the MLE method does not always generate something that looks like a least squares. For example, in a radioactive decay experiment the statistics of the measurements are expected to be Poisson. If the average (or expected) number of counts in a time interval is λ, then the probability of m counts is given by

$$p(m) = \exp(-\lambda)\frac{\lambda^m}{m!}\,.$$

Using our model function, we see that in the nth interval, we expect $f(n, \alpha)$ counts. In other words,

$$\lambda = f(n, \alpha)\,.$$

If we measure d_n counts in that interval, the probability of that measurement given the set of parameters α is given by

$$p(d_n|\boldsymbol{\alpha}) = e^{-f(n,\boldsymbol{\alpha})}\frac{f(n, \boldsymbol{\alpha})^{d_n}}{d_n!}$$

$$\log(p(\{d_n\}|\boldsymbol{\alpha})) = -\sum_n f(n, \boldsymbol{\alpha}) + \sum_n d_n\, f(n, \boldsymbol{\alpha}) - \sum_n \log(d_n!)\,. \quad (4.2.8)$$

The MLE equations are derived by taking the derivative with respect to α_i of the above expression for the total probability of the data set. Note that d_n is not a function of $\boldsymbol{\alpha}$. The equations derived for this are

$$\sum_n \left(\frac{d_n}{f(n, \boldsymbol{\alpha})} - 1\right)\frac{\partial f(n, \boldsymbol{\alpha})}{\partial \alpha_i} = 0, \quad i = 0, 1, \dots \quad (4.2.9)$$

$$f(n, \boldsymbol{\alpha}) = \alpha_0 \exp(-\alpha_1 n), \quad n = 0.5, 1.5, 2.5, 3.5\dots$$

This is clearly not a least squares problem. Note also that because of the appearance of the model function in the denominator of the MLE equation for a Poisson process, the data points with the smallest expected value are weighted the strongest.

Example.   The table below is a set of numbers generated assuming that $f(n, \boldsymbol{\alpha}) = 100\exp[-0.3n]$. Eleven data points were generated for $n = 0.5, 1.5, 2.5\dots$. These numbers were then used as input to a Poisson generator in Mathematica®. The table is the output of the Poisson generator.

Table 4.2

96	65	47	36	26	18	8	3	6	3
2	5	4	2	2	1	3	0	1	

Using Equations 4.2.9 on this data set, I got the result

$$\alpha_0 = 101.1 \qquad \alpha_1 = 0.285 \,.$$

Try getting a solution using the log of the data and then again using 4.2.9.

4.2.3 *Non-independent Gaussian statistics*

It can happen that measurements are not independent. As an example, assume that the pdf for a set of N measurements in a stationary system have a joint Gaussian distribution. In this case,

$$p(d_1, d_2, \ldots, d_N | \boldsymbol{\alpha}) = (2\pi)^{-N/2} |\boldsymbol{\Lambda}|^{-1/2} \exp(-0.5(\Delta d) \cdot \boldsymbol{\Lambda}^{-1} \cdot \Delta d) \,, \quad (4.2.10)$$

where d is the vector of random variables,

$$\Delta d_n = d_n - f(n, \boldsymbol{\alpha})$$

$$\Lambda_{nm} = \langle \Delta d_n \Delta d_m \rangle \,,$$

and $|\boldsymbol{\Lambda}|$ is the determinant of $\boldsymbol{\Lambda}$.

$$\log p(\{d_n\} | ?) = -\frac{1}{2} \sum_n \sum_m \Lambda_{nm}^{-1} \Delta d_n \Delta d_m - \frac{\log |\boldsymbol{\Lambda}|}{2} - \frac{N \log 2\pi}{2} \quad (4.2.11)$$

$$\Lambda_{nm} = \langle \Delta d_n \Delta d_m \rangle = \sigma^2 R(n - m) \,. \quad (4.2.12)$$

Again, specification of the pdf completes the specification of the problem.

Example: There is a wonderfully clever example of dealing with a correlated time series published by Aleksey Lomakin. Consider a zero mean time series $\{d_n\}$ whose expected autocorrelation function is $\alpha_0 \exp[-\alpha_1 n]$. The probability for the data set given the parameters is 4.2.10 and the log is given by 4.2.11. The exponent term is

$$-\frac{1}{2} \sum_{k=0}^{N-1} \sum_{j=0}^{N-1} d_k \Lambda_{kj}^{-1} d_j \,.$$

The sums run over all the data points in the measured time series. The term Λ_{jk}^{-1} is the inverse matrix of the expected autocorrelation function of the time series. As such it is only a function of $|k - j|$. If the sum over j is done with $(k - j)$ constant, this expression can be rearranged to get

$$-\frac{1}{2} \sum_{k-j=-(N-1)}^{N-1} \Lambda_{jk}^{-1} \sum_{j=0}^{N-|k-j|-1} d_j d_{j+(k-j)} \,.$$

The inner sum is the un-normalized experimental autocorrelation function, *viz.* (see 3.5.1)

$$\sum_{j=0}^{N-|k-j|-1} d_j d_{j+(k-j)} = N \times RC(k - j) \,.$$

So, to terms of the order $1/N$, the exponential term can be written

$$-\frac{N}{2} \sum_{m=-N}^{N} RC(m) \Lambda_m^{-1} \,. \tag{4.2.13}$$

In this situation, the matrix Λ^{-1} is a symmetric Toeplitz matrix and as such has a special form, *viz.*

$$\Lambda_{jk} = R(j - k) = \alpha_0 \exp[-\alpha_1 |j - k|] = \alpha_0 a^{|j-k|} \,,$$

where

$$a = \exp[-\alpha_1] \,.$$

Then, it can be shown that

$$\det(\Lambda) = \alpha_0^N (1 - a^2)^{N-1}$$

and the inverse matrix is tri-diagonal, *viz.*

$$\Lambda_{11}^{-1} = \Lambda_{NN}^{-1} = \frac{1}{\alpha_0(1 - a^2)} \tag{4.2.14}$$

$$\Lambda_{kk}^{-1} = \frac{(1 + a^2)}{\alpha_0(1 - a^2)}, \quad k \neq 1, N$$

$$\Lambda_{k(k-1)}^{-1} = \frac{-a}{\alpha_0(1 - a^2)}, \quad k = 2 \ldots N$$

$$\Lambda_{k(k+1)}^{-1} = \frac{-a}{\alpha_0(1 - a^2)}, \quad k = 1 \ldots N - 1$$

$$\Lambda_{kj}^{-1} = 0 \quad \text{Otherwise.}$$

Note that the only "m" ($= j - k$) values in Equation 4.1.15 that are non-zero are 0, -1, and 1. In the limit of large N, the log probability can thus be written

$$-\frac{N}{2}\left[\frac{RC[0](1 + e^{-2\alpha_1}) + 2e^{-\alpha_1}RC[1]}{\alpha_0(1 - e^{-2\alpha_1})}\right] - \frac{N}{2}\ln\alpha_0 - \frac{N}{2}\ln(1 - e^{-2\alpha_1}) + \text{const.}$$

This probability is maximized when

$$\alpha_0 = RC[0] \quad and \quad \alpha_1 = -\ln\frac{RC[0]}{RC[1]}.$$

The optimal result is obtained using only 2 points of the measured correlogram. This result was a surprise to me, but I have checked the arithmetic very carefully and it appears to be correct. Perhaps it should not be too much of a surprise since $RC[0]$ and $RC[1]$ are computed using all the data points.

In general, estimations involving correlation functions are more complicated, but the principles outlined in this section will still have to be followed.

4.2.4 *Chi-squared distributions*

Recall that in Chapter 2 we did an exercise to illustrate the variance of the mean. We did an experiment where we used 100 examples to estimate the variance of mean for means taken using different numbers of samples. The examples were for means measured using 10, 20, 50 and 100 points. The corresponding measured variances were 0.00811, 0.00322, 0.00149, and 0.00067.

Could we use the method discussed in this chapter to estimate the coefficients for an expression of the form

$$\sigma_N^2 = AN^q?$$

In this case, we are using 100 points to estimate the variances, so we can use the asymptotic form of the appropriate distribution given by 1.7.24, *viz.*

$$p(s_N^2) = \frac{1}{\sqrt{4\pi(\sigma_N^2)^2/(\nu - 1)}}\text{Exp}\left[-\frac{(s_N^2 - \sigma_N^2)^2}{4(\sigma_N^2)^2/(\nu - 1)}\right].$$

Here, s_N^2 is the measured variance. The logarithm of the probability of the data set, given the model is

$$\ln[p] = -\sum_N \frac{(s_N^2 - AN^q)^2}{4(AN^q)^2/\nu - 1} - \sum_N \ln[AN^q] + \text{const.}$$

When I maximize this probability by varying N and p, I get

$$A = 0.0855 \quad \text{and} \quad q = -1.055\,.$$

The theoretical values for these parameters are

$$A = 0.08333 \quad \text{and} \quad q = -1\,.$$

4.3   Residuals

A question that often arises is how you tell if you used the correct model function and statistics? That question can be partly answered by examining the so-called weighted residuals. Let

$$w_n = \frac{(d_n - f(n, \boldsymbol{\alpha}))}{\sigma(n, \boldsymbol{\alpha})}\,. \tag{4.3.1}$$

It is the difference between your measured data point and the model function weighted by the function $\sigma(n, \boldsymbol{\alpha})$, the expected standard deviation at that point. The calculation of the standard deviation changes depending on what kind of statistics you think you have.

If the statistics are Poisson,

$$\sigma = \sqrt{f(n, \boldsymbol{\alpha})}\,. \tag{4.3.2}$$

Now let's compute the expected value and variance of w_n. We do it first for the Gaussian case.

$$\langle w_n \rangle = \langle \frac{d_n - f(n, \boldsymbol{\alpha})}{\sigma} \rangle = 0$$

$$\langle w_n^2 \rangle = \frac{\langle (d_n - f(n, \boldsymbol{\alpha}))^2 \rangle}{\sigma^2} = 1\,.$$

The mean is zero and the expected variance is 1.

Likewise, for the Poisson distribution,

$$\langle w_n \rangle = \left\langle \frac{d_n - f(n, \boldsymbol{\alpha})}{\sqrt{f(n, \sigma)}} \right\rangle = 0\,,$$

$$\langle w_n^2 \langle = \left\langle \frac{(d_n - f(n, \boldsymbol{\alpha}))^2}{f(n, \sigma)} \right\rangle$$

$$= \frac{1}{f}\langle (d_n - f(n, \boldsymbol{\alpha}))^2 \rangle = \frac{1}{f}(\langle d_n^2 \rangle - f^2).$$

But for a Poisson distribution,

$$\langle d_n^2 \rangle = f^2 + f,$$

so

$$\langle w_n^2 \rangle = 1.$$

The weighted residuals should be a set of random numbers with an expected mean of zero and an expected variance of one. Further, since a critical assumption in the analysis to this point is that the residuals are independent of each other, the weighted residuals should also be independent of each other. A plot of the weighted residuals should appear to be featureless. There should be no apparent pattern in it. If there is a pattern in the residuals plot or if the mean or variance are incorrect, then the model function and/or the statistics are wrong. If there is a positive or negative trend in the weighted residuals, the model is probably wrong. If there is no trend in the running mean of residuals, but the size of the residuals is varying systematically, the statistical model is probably wrong.

Figure 4.2 shows a plot of the residuals for the heat transfer example where the fit was done using a non-linear least squares (MLE) procedure. Note the lack of an apparent pattern. However, note also that the variance is not 1. It is fairly typical when you are doing a problem with Gaussian statistics that you don't know the variance. All you know is that the variance is constant. In that case, you get the same information about the quality of the fit from a weighted residuals plot as you do from an unweighted one. In fact, the variance can be estimated from the residuals data. In this example, it was calculated to be 14.4.

Figure4.3 is a plot of the residuals for the same temperature data, but using the fit obtained by taking the logarithm of the data. Notice how large the residuals are at the beginning of the data set. This is fairly typical of fits made to the logarithm of the data. As predicted, the fit tended to emphasize the smaller values in the data. This residuals plot obviously has structure in it. The assumed statistics were wrong, even though we had the model function correct.

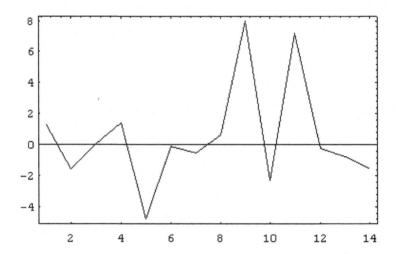

Fig. 4.2   Residuals for non-linear fit to exponential.

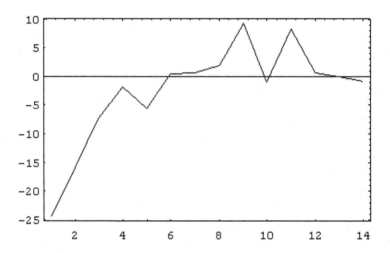

Fig. 4.3   Residuals for a fit to the logarithm of exponential.

4.4   Parameter Error Estimates

A question that is not asked enough is how good the estimates of the parameters α are. Until now, we have merely explained how to get the best estimates of the parameters. We have given no hint as to how good the estimates are. We should always keep in mind that the procedure for deriving the parameters is really an elaborate procedure for making a new random variable. If you do an identical parameter estimation scheme with two different sets of data from the same experiment, I guarantee that you will get two different sets of answers. This section is devoted to deriving expressions that give an estimate of the variation in the answers.

The method for getting the error estimates is based on examining the sharpness of the peak in the log of the probability of the data set, as a function of α_i. If the peak is very sharp, the error in the estimation of α_i is expected to be small. Likewise, if the peak is broad, the error in the parameter estimate is expected to be large. To make this more concrete, I will do a simple example of a least squares problem with independent errors with a constant variance, σ^2, and one parameter. The argument is easily extended to the multiple parameter case.

Consider a model function, f. At the "correct" value of the parameter, denote the function, $f_{0,n}$. In any particular realization of the experiment, a different parameter value is obtained, the error. Let $\Delta\alpha$ be the error for the experiment. Near the correct value of α, the function f can be written

$$f(n,\alpha) = f(n,\alpha_0) + \left.\frac{\partial f(n,\alpha)}{\partial \alpha}\right|_{\alpha_0} (\alpha - \alpha_0) = f_{n,0} + \frac{\partial f_{n,0}}{\partial \alpha}\Delta\alpha.$$

In this case, the sum of squares can be written

$$\varepsilon^2 = \sum_n \left(f_{0,n} + \frac{\partial f_{0,n}}{\partial \alpha}\Delta\alpha - d_n \right)^2,$$

where d_n are the data points.

The solution to the least squares problem is obtained, when

$$\frac{\partial \varepsilon^2}{\partial \alpha} = 0 = 2\sum_n \frac{\partial f_{0,n}}{\partial \alpha}\left(f_{0,n} + \frac{\partial f_{0,n}}{\partial \alpha}\Delta\alpha - d_n \right). \tag{4.4.1}$$

Manipulating this equation, you get

$$\Delta\alpha = -\frac{\displaystyle\sum_n \frac{\partial f_{0,n}}{\partial \alpha}(f_{0,n} - d_n)}{\displaystyle\sum_n \left(\frac{\partial f_{0,n}}{\partial \alpha} \right)^2}.$$

The expected value of the error in the parameter estimate is zero. We can use this expression to estimate the variance of the error.

$$\Delta\alpha^2 = \frac{\sum_n \sum_m \dfrac{\partial f_{0,n}}{\partial\alpha} \dfrac{\partial f_{0,m}}{\partial\alpha}(f_{0,n} - d_n)(f_{0,m} - d_m)}{\left(\sum_n \left(\dfrac{\partial f_{0,n}}{\partial\alpha}\right)^2\right)^2}.$$

If we take the expected value of this expression, noting that the errors in each data point are independent, we get

$$\langle\Delta\alpha^2\rangle = \frac{\sigma^2 \sum_n \left(\dfrac{\partial f_{0,n}}{\partial\alpha}\right)^2}{\left(\sum_n \left(\dfrac{\partial f_{0,n}}{\partial\alpha}\right)^2\right)^2} = \frac{\sigma^2}{\sum_n \left(\dfrac{\partial f_{0,n}}{\partial\alpha}\right)^2}. \qquad (4.4.2)$$

Note from 4.4.1, it can be seen that

$$\left\langle\frac{\partial^2\varepsilon^2}{\partial\alpha^2}\right\rangle = 2\sum_n \left(\frac{\partial f_{0,n}}{\partial\alpha}\right)^2.$$

Thus, 4.4.2 can be rewritten

$$\langle\Delta\alpha^2\rangle = \frac{2\sigma^2}{\left\langle\dfrac{\partial^2\varepsilon^2}{\partial\alpha^2}\right\rangle}.$$

Recalling that for this Gaussian case,

$$\ln[p] \approx -\frac{\varepsilon^2}{2\sigma^2}.$$

Thus,

$$\langle\Delta\alpha^2\rangle = \left(-\frac{\partial^2 \ln p}{\partial\alpha^2}\right)^{-1}.$$

The error in the parameter estimate can now be seen to be inversely proportional to minus the second derivative of the log of the probability of the data set, given the parameter.

This concept will be expanded below. The measure of the sharpness of the peak used is the second derivative of the log of the probability taken with respect to the estimated parameter. The procedure is as follows:

Compute the matrix F_{ij},

$$F_{ij} = \left\langle -\frac{\partial^2 \log p(\{d_n\}|\boldsymbol{\alpha})}{\partial \alpha_i \partial \alpha_i} \right\rangle, \tag{4.4.3}$$

where $\langle \rangle$ denotes expected value. The lower bound on the estimate of the parameters (called the *Cramer–Rao bound*) is given by

$$\sigma_i^2 \geq F_{ii}^{-1}. \tag{4.4.4}$$

This places a lower limit on the error estimate. It would be nice if it placed an upper limit. This expression becomes an equality, if the statistics of the original problem are Gaussian. However, my experience has been that this bound is a useful estimate of the measurement error, *i.e.*

$$\sigma_1^2 \approx F_{ii}^{-1}. \tag{4.4.5}$$

F_{ii}^{-1} is the ith diagonal element of the matrix inverse of F_{ij}.
The matrix $\boldsymbol{F}$ is referred to as the Fisher matrix.

4.4.1 Expected errors for Gaussian statistics

Recall that for a Gaussian

$$\log p(\{d_n\}|\boldsymbol{\alpha}) = -\sum_n \frac{(d_n - f(n, \boldsymbol{\alpha}))^2}{2\sigma^2} - \frac{1}{2} \sum_n \log \ 2\pi\sigma^2.$$

Thus

$$\frac{\partial \log p(\{d_n\}|\boldsymbol{\alpha})}{\partial \alpha_i} = \sum_n \frac{1}{\sigma^2} \frac{\partial f_n}{\partial \alpha_i} (d_n - f) + \frac{(d_n - f_n)^2}{2(\sigma^2)^2} \frac{\partial \sigma^2}{\partial \alpha_i} - 0.5 \frac{\partial \sigma^2}{\sigma^2 \partial \alpha_i}$$

$$\frac{\partial^2 \log p(\{d_n\}|\boldsymbol{\alpha})}{\partial \alpha_i \partial \alpha_j} = \sum_n \frac{1}{\sigma^2} \frac{\partial^2 f_n}{\partial \alpha_i \partial \alpha_j} (d_n - f) - \frac{1}{\sigma^2} \frac{\partial f_n}{\partial \alpha_i} \frac{\partial f_n}{\partial \alpha_j}$$

$$- \frac{(d_n - f_n)^2}{(\sigma^2)^2} \frac{\partial f_n}{\partial \alpha_i} \frac{\partial \sigma^2}{\partial \alpha_j} - \frac{(d_n - f_n)}{(\sigma^2)^2} \frac{\partial f_n}{\partial \alpha_j} \frac{\partial \sigma^2}{\partial \alpha_i}$$

$$- \frac{(d_n - f_n)^2}{(\sigma^2)^3} \frac{\partial \sigma^2}{\partial \alpha_i} \frac{\partial \sigma^2}{\partial \alpha_j} + \frac{(d_n - f_n)^2}{2(\sigma^2)^2} \frac{\partial^2 \sigma^2}{\partial \alpha_i \partial \alpha_j}$$

$$- \frac{1}{2} \frac{1}{\sigma^2} \frac{\partial^2 \sigma^2}{\partial \alpha_i \partial \alpha_j} + \frac{1}{2} \frac{1}{(\sigma^2)^2} \frac{\partial \sigma^2}{\partial \alpha_i} \frac{\partial \sigma^2}{\partial \alpha_j}.$$

Fortunately, this formidable expression simplifies greatly when we take the expected value. Recalling that

$$\langle (d_n - f_n) \rangle = 0,$$

and

$$\langle (d_n - f_n)^2 \rangle = \sigma_n^2,$$

we get

$$\left\langle -\frac{\partial^2 \log p(\{d_n\}|\boldsymbol{\alpha})}{\partial \alpha_i \partial \alpha_j} \right\rangle = \sum_n \frac{1}{\sigma_n^2} \frac{\partial f_n}{\partial \alpha_i} \frac{\partial f_n}{\partial \alpha_j} + \frac{1}{2(\sigma_n^2)^2} \frac{\partial \sigma_n^2}{\partial \alpha_i} \frac{\partial \sigma_n^2}{\partial \alpha_j}. \qquad (4.4.6)$$

For the heat transfer problem demonstrated here,

$$F_{00} = \frac{1}{\sigma^2} \sum \exp(-2\alpha_1 n),$$

$$F_{01} = F_{10} = -\frac{1}{\sigma^2} \sum \alpha_0 n \exp(-2\alpha_1 n),$$

$$F_{11} = \frac{1}{\sigma^2} \sum \alpha_0^2 n^2 \exp(-2\alpha_1 n).$$

In order to evaluate this expression, you use the values for α obtained in the parameter estimation process. Recall that the variance was not known at the beginning of the fitting process, although we did know that the statistics were Gaussian. The measured variance was 14.4. The values of F_{ij} are found to be

$$F_{00} = 0.155, F_{01} = F_{10} = -18.25, F_{11} = 6152.$$

The inverse matrix containing the error estimates is thus,

$$F_{00}^{-1} = 9.948, F_{01}^{-1} = F_{10}^{-1} = 0.2951, F_{11}^{-1} = 2.5x10^{-4}.$$

The computed error estimates of the estimation procedure are

$$\sigma_{\alpha_0} \approx 3.15, \text{ and } \sigma_{\alpha_1} \approx 0.016.$$

The computed numbers are the estimated standard deviation of the parameters derived. There is an implied assumption here that *the statistics of the estimates* are Gaussian. Given that, there is a 67% chance that the expected value lies within $\pm \sigma$ of the derived parameter. If you then report the results of your experiment in the form $\alpha_i \pm \sigma_i$, the reader will know not only your result, but the 67% confidence limits.

The off diagonal terms in the matrix are the cross correlation coefficients. They are a measure of how the error in the ith parameter estimate affects the error in the jth parameter estimate.

In the case that you are using the correct Gaussian statistics and the wrong model, it can be shown that the Fisher matrix gives you an estimate of the reproducibility error. In other words, if you repeat the experiment with the same incorrect model, the Fisher matrix gives you information about how close the estimated parameters will be to the last experiment.

4.4.2 *Error estimates for joint Gaussian statistics*

For a joint Gaussian distribution, the Fisher matrix calculation simplifies to

$$F_{ij} = \sum_p \left(\sum_l \Lambda_{pl}^{-1} \sum_m \frac{\partial \Lambda_{lm}}{\partial \alpha_i} \sum_k \Lambda_{mk}^{-1} \frac{\partial \Lambda_{kp}}{\partial \alpha_j} \right). \qquad (4.4.7)$$

In the example in Section 4.2.3, it is possible to compute the Fisher matrix analytically. Using Mathematica®, I got

$$F_{00} = \frac{(7 + e^{2\alpha_1})N}{2\alpha_0^2(e^{2\alpha_1} - 1)}, \quad F_{01} = F_{10} = \frac{(1 + 3e^{2\alpha_1})N}{\alpha_0(e^{2\alpha_1} - 1)^2}$$

$$F_{11} = \frac{N(1 + 12e^{2\alpha_1} + 3e^{4\alpha_1})}{(e^{2\alpha_1} - 1)^3}.$$

Usually the exponential decay term is what you are interested in so I will just compute the estimated error for it.

Taking the inverse to get the Cramer–Rao bound for α_1, I get an estimate of the relative error for the exponential term

$$\frac{\sigma_{\alpha_1}}{\alpha_1} = \frac{1}{\alpha_1} \sqrt{\frac{(e^{2\alpha_1} - 1)^3(e^{2\alpha_1} + 7)}{N(3e^{6\alpha_1} + 15e^{4\alpha_1} + 73e^{2\alpha_1} + 5)}}.$$

For $a_1 = 0.3$,

$$\frac{\sigma_{\alpha_1}}{\alpha_1} \approx \frac{0.51}{\sqrt{N}}.$$

Using $N = 1000$, not many points for a measured correlogram, the relative error is only 1.6%.

4.4.3 *Error estimates for Poisson processes*

For a Poisson process, it can easily be shown that

$$F_{ij} = \sum_n \frac{1}{f(n, \alpha)} \frac{\partial f}{\partial \alpha_i} \frac{\partial f}{\partial \alpha_j}. \tag{4.4.8}$$

For the example problem in 4.1.2,

$$F_{00} = \sum_n \frac{\mathrm{Exp}(-\alpha_1 n)}{\alpha_0}, \qquad F_{01} = F_{10} = -\sum_n n\mathrm{Exp}(-\alpha_1 n)$$

$$F_{11} = \sum_n n^2 \alpha_0 \mathrm{Exp}(-\alpha_1 n)$$

For the example here, $F_{00} = 0.032$, $F_{01} = F_{10} = -9.39$, $F_{11} = 4747.4$. The inverse matrix gives

$$\sigma_{\alpha_0} \geq 8.64 \text{ and } \sigma_{\alpha_1} \geq 0.02 .$$

Note that this error is larger than for the Gaussian example, using the same parameters.

4.4.4 *Error estimates for Chi-square distributions*

The Fisher matrix for the example in Section 4.2.4 takes on a particularly simple form, *viz.*

$$F_{00} = \sum_N \frac{\nu - 3}{2A^2}, \quad F_{01} = F_{10} = \frac{\nu - 1}{2A} \sum_N \ln[N], \quad F_{11} = \frac{\nu - 1}{2} \sum_N \ln[N]^2 .$$

For the example problem

$$F_{00} = 25600, \quad F_{01} = F_{10} = 7855.8, \quad F_{11} = 2514 ,$$

which gives the results

$$\sigma_A \geq 0.031, \quad \sigma_p \geq 0.098 .$$

This result is entirely consistent with the theoretical result.

4.5 *A Priori Error Estimation*

Note that data is not needed to evaluate the error estimates. All you need is some pre-knowledge about approximately what value the parameters may have. It is always true that you have some idea of the range of the possible answers from an experiment. If you claim to have no idea of what the possible answers are, you don't have any idea of the size and range of the instruments you need to do the experiment. Besides, computer experiments are cheap. You can check the error estimates for large ranges of possible answers.

The point is that *the MLE error estimates can be used ahead of time to set up the optimal experimental conditions.* As an example, suppose you were able to take only 4 equally spaced data points in the heat transfer experiment. Where should they be to give the best possible results?

The procedure is to evaluate the F matrix and its inverse using the approximate values of the parameters and varying the placement of the data points to find the placement that will give the smallest variance in the most important parameter. In this particular experiment, we were interested in the heat transfer coefficient, so we want to minimize the measurement variance of α_1.

One way to proceed here is to define a time "stretch" factor β. The model function is thus written

$$f(n, \boldsymbol{\alpha}) = \alpha_0 \exp(-\alpha_1 \beta n).$$

Time is stretched by the factor β. The equations for the Fisher Matrix are now

$$F_{00} = \frac{1}{\sigma^2} \sum \exp(-2\alpha_1 \beta n), \ F_{01} = F_{10} = -\frac{\beta}{\sigma^2} \sum \alpha_0 \ n \ \exp(-2\alpha_1 \beta n),$$

$$F_{11} = \frac{\beta^2}{\sigma^2} \sum \alpha_0^2 n^2 \exp(-2\alpha_1 \beta n).$$

Figure 4.4 is a plot of the expected standard deviation in the estimate of α_1 as a function of the stretch factor. Note that there is a minimum in the error at a stretch factor of approximately 1.9. This gives a minimum expected error of.0.023. Not bad for only 4 measurements. Recall that before we had an expected error of.0.016 with 20 measurements.

An interesting exercise is to do the same thing for the radioactive decay experiment. Here the statistics are Poisson, so we do not expect to see the same calculations. The equations for the Fisher matrix (4.1.8) are:

$$F_{00} = \frac{1}{\alpha_0} \sum \exp(-\alpha_1 \beta n), \ F_{01} = F_{10} = -\beta \sum n \exp(-\alpha_1 \beta n),$$

$$F_{11} = \beta^2 \sum \alpha_0 \ n^2 \exp(-\alpha_1 \beta \ n).$$

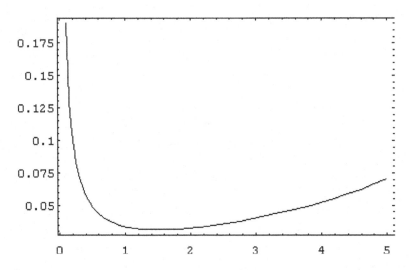

Fig. 4.4 σ_{α_1} vs. stretch factor β for the example with Gaussian statistics.

In this example, $n = 0.5, 1.5, 2.5, 3.5$.

Figure 4.5 again shows a plot of the expected standard deviation of the measurement error as a function of the stretch parameter β. Note that the optimal β here is close to 2.9 and the minimum error is 0.04. That is definitely different from what we got for the experiment that had Gaussian statistics. The optimal stretch factor is a third again as large. This partly reflects the fact that Poisson statistics MLE techniques tend to emphasize the regions where the data is expected to be smallest.

4.6   Maximum *A Posteriori* Estimation

In this section, we consider the effect of some *a priori* knowledge of the parameter set α. Denote the pdf for α, $p(\alpha)$. Then, from Section 2.1,

$$p(\alpha, \{d_i\}) = p(\{d_i\}|\alpha)p(\alpha)/p(\{d_i\}). \tag{4.6.1}$$

The MAP equations are

$$\frac{\partial \log p(\alpha|\{d_n\})}{\partial a_i} = 0, \qquad i = 1, 2, 3, \ldots \tag{4.6.2}$$

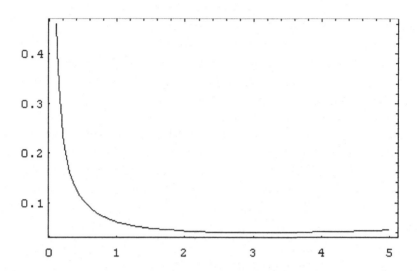

Fig. 4.5 σ_{α_1} vs. stretch factor β — Poisson statistics.

where

$$\log(p(\boldsymbol{\alpha}|\{d_n\})) = \log(p(\{d_n\}|\boldsymbol{\alpha}) + \log(p(\boldsymbol{\alpha})) - \log(p(d_i))\,.$$

As before, $\{d_i\}$ is not a function of $\boldsymbol{\alpha}$, so the probability for this term is not carried in the calculations. If $p(\boldsymbol{\alpha})$ represents simple pre-knowledge about the allowed range of the parameters, the probability is a uniform probability and has no effect on the steps involving derivatives of the log probabilities. On the other hand, if $\log p(\boldsymbol{\alpha})$ is a known, differentiable function, it can be carried in the solution procedure.

For example, suppose $p(\alpha)$ is a Gaussian for one of the variables, say α_1. Then

$$\log\ p(\alpha) = \frac{-(\alpha_1 - \langle\alpha_1\rangle)^2}{2\sigma_{\alpha_1}^2} - 0.5\log 2\pi\sigma_{\alpha_1}^2\,.$$

In this case, the Poisson statistical estimation problem of Section 4.1.2 would become

$$\sum_n \left(\frac{d_n}{f(n,\alpha)} - 1\right)\frac{\partial f(n,\alpha)}{\partial\alpha_i} + \frac{(\alpha_1 - \langle\alpha_1\rangle)}{\sigma_{\alpha_1}^2}\delta_{1i} = 0, i = 0, 1\,. \qquad (4.6.3)$$

I have implicitly assumed that σ_{α_1} is not a function of α_1.

The Fisher matrix in this case becomes

$$F_{ij} = \left\langle -\frac{\partial^2 p(\alpha|\{d_n\})}{\partial\alpha_i\partial\alpha_j} \right\rangle = \left\langle -\frac{\partial^2 \log p(\{d_n\}|\alpha)}{\partial\alpha_i\partial\alpha_j} \right\rangle + \frac{\delta_{1i}\delta_{1j}}{\sigma^2_{\alpha_1}}.$$

As before, the inverse Fisher matrix gives the estimates of the measurement errors. There is an additional term to the expression for the MLE Fisher matrix for the diagonal term involving the estimate of α_1. This term is positive, so the effect is to decrease the magnitude of the terms in the inverse matrix, i.e., decreasing the estimate of the measurement error. The better you know the parameter α_i, the better your estimate of it from the data set. Or, the more you know, the more you can know.

There are several interesting philosophical aspect of MAP problems. Not the least of which is dealing with the concept that you already have $\langle\alpha\rangle$. If you already know the expected value, why bother to estimate it? If you really are doing an experiment with a random variable, at best, you can have some estimate of $\langle\alpha\rangle$. This estimate could have come from a previous experiment or from theory. As a practical matter, using MAP with a previous estimate of $\langle\alpha\rangle$ decreases the range of possible outcomes of the solution procedure and often speeds up the computations.

On the other hand, it is common that you have a situation where you have pre-knowledge of at least one of the parameters and want to include that information in the estimation procedure. The nature of MLE and MAP procedures is that all the unknown parameters must be estimated, even if only a subset are really of interest. Auxiliary parameters are those parameters that must be estimated but are not the primary focus of the experiment. For instance, assume you are doing an experiment trying to estimate the decay constant of an exponential, but where there is a small random baseline, generated by your measurement system, *viz.*

$$f(n, \boldsymbol{\alpha}) = 10e^{-\alpha_0 n} + \alpha_1.$$

The parameter you are interested in is α_0, and α_1 is the auxiliary parameter. The point is that you have to estimate α_1, even though you have a good idea of what it is, but you do not have exact pre-knowledge. If you used a MLE procedure, assuming the most likely value of the baseline was zero, the Fisher matrix would be

$$F_{00} = \sum_{n=1}^{N} n^2 e^{-\alpha_0 n}; F_{01} = F_{10} = -N(N+1)/2;$$

$$F_{11} = \sum_{n=1}^{N} e^{\alpha_0 n}.$$

If you used a MAP procedure, the Fisher matrix would be

$$F_{00} = \sum_{n=1}^{N} n^2 e^{-\alpha_0 n}; F_{01} = F_{10} = -N(N+1)/2;$$

$$F_{11} = \sum_{n=1}^{N} e^{\alpha_0 n} + \frac{1}{\sigma_{\alpha_1}^2}.$$

The F_{11} term is increased over the MLE case. To illustrate the effect, assume $\alpha_0 \approx 0.5$. Further, assume you are using 10 data points in the procedure and you are sure the baseline is zero, ± 0.05.

In the MLE case, you get

$$F_{00} = 143.2; F_{01} = F_{10} = 55; F_{11} = 37.465.$$

The inverse matrix J is

$$J_{00} = 0.016; J_{01} = J_{10} = -0.0235, J_{11} = 0.0612.$$

The estimated measurement error of α_0 would be approximately 0.1265. For the MAP, using the pre-knowledge about α_0, you get

$$F_{00} = 143.2; F_{01} = F_{10} = 55; F_{11} = 437.465.$$

The inverse matrix J is

$$J_{00} = 0.007; J_{01} = J_{10} = -0.0009, J_{11} = 0.0024,$$

for an error estimate on α_0 of 0.0857, adding the pre-knowledge made for a substantial decrease in the expected measurement error.

It is also instructive to look at the off-diagonal terms in the inverse of the Fisher matrix. These terms give you information about the coupling between the error estimates of the parameters. Form the terms

$$\varepsilon_{ij} = \frac{J_{ij}}{\sqrt{J_{ii}J_{jj}}}. \tag{4.6.4}$$

The ϵ_{ij} terms may be viewed as correlation coefficients between the errors in the estimation of the parameters. For instance, in the last example, the correlation coefficient was -0.751 for the MLE case, and -0.22, in the MAP case. In the MLE case, the measurement errors are strongly coupled. The calculation indicates that if the estimation in α_1 is low, the estimation in α_0 is likely to be high. The trend is the same in the MAP case, but the coupling between the measurement errors is much smaller. In fact, if you examine this problem closely, you will see that as the pre-knowledge in α_1 increases, the main effect is to decrease the coupling in the errors. In other words, J_{00} gets closer to $1/F_{00}$.

References

1. "Fitting the Correlation Function," Aleksey Lomakin, Applied Optics, Vol. 40, No. 24, 20 August 2001, pp. 4079–4086.
2. "Detection, Estimation, and Modulation Theory," Harry L.Van Trees, Part I, John Wiley and Sons, 1968.

Chapter 5

Random Sampling

5.1 Basic Concepts of Random Sampling

For various reasons, time series are often encountered that are created by random sampling. That is, the samples are not taken at regular time intervals, but are instead taken at random times. The rate of sampling may or may not be a function of the variable being sampled.

I will consider the two situations separately: namely, random sampling independent of the measured variable, and random sampling that depends on the variable being sampled.

In a situation like this, where there are two random variables involved, we will need the pdf for both random variables in order to do the expected value calculations correctly.

Let f_i represent the ith sample of a random variable, f, measured at time t_i.

In general, the probability of a measurement of f at time t_i is given by a joint probability

$$p(f, t_i).$$

Invoking Bayes' rule, we may write

$$p(f, t_i) = p(t_i|f)p(f).$$ (5.1.1)

This form emphasizes the possible dependence of the measurement probability on the value of the measurement f. The term $p(t_i|f)$ can be seen to be the relative measurement rate for value f. More on this later.

5.2　Independent Sampling

In this section, I will assume sampling independent of the time series f. The time series f is assumed to have an autocorrelation function and associated correlation times. In this section, we will examine how random sampling affects our ability to extract information from the time series. Since the probability of measurement is independent of the value, can write

$$p(t_i, f) = p(t_i)p(f) \,.$$

First, consider the statistics of the sample times, t_i. There are many possibilities, but I will concentrate on the case where the sampling times are completely independent of each other. This situation can come about when one is measuring velocities by looking at the speed of particles entrained in the flow. You get a measurement when a particle goes through the pre-set measurement region. It is not too hard to see that the probability of a measurement in the infinitesimal interval (t, dt) is given by

$$p(t_i)dt_i = \rho dt_i \,,$$

where ρ is the expected number of measurements per unit time. But, since

$$1 = \int_{\text{all } t} p(t)dt \,,$$

it can seen that for any finite interval T sufficiently large, the probability of a measurement at time t, dt is given by

$$p(t)dt = \frac{dt}{T} \,. \tag{5.2.1}$$

This expresses the concept that any time in the interval is as likely as any other interval.

There are many possibilities for models for random sampling. Here, I will use the so-called totally random distribution, the Poisson distribution. The probability for n measurements in a finite time interval, Δt, is given by

$$p(n) = e^{-\rho \Delta t} \frac{(\rho \Delta t)^n}{n!} \,. \tag{5.2.2}$$

The statistics of the time between measurements, ΔT, *the inter-arrival time distribution* can be written,

$$p(\Delta T)d\Delta T = e^{-\rho\Delta T}\rho d\Delta T, \qquad (5.2.3)$$

from which it can be seen that the average time between measurements is $1/\rho$.

Armed with the above information, we can compute the expected mean and measurement variance of the randomly sampled data.

5.2.1 *Mean and variance of the mean with random sampling*

$$\langle \bar{f} \rangle = \left\langle \frac{1}{N} \sum_k f(t_k) \right\rangle$$

$$= \frac{1}{N} \sum_k \langle f(t_k) \rangle.$$

$$\langle \bar{f} \rangle = \frac{1}{N} \sum_k \langle f(t_k) \rangle = \int_T \int_f p(t)p(f)f\,dt\,df = \int_f p(f)f\,df = \langle f \rangle.$$

The expected value of the time series is still the same as for a regularly sampled series.

We can compute the expected measurement variance for a randomly sampled mean using N points. Before doing the detailed calculations, it is useful to examine the expression for the measurement variance and try to relate to similar calculations we have done before. If the measurement rate, ρ, is such that there are few measurements per integral correlation time, the measurements are essentially independent and we would expect the variance of the mean, σ_N^2, to be

$$\sigma_N^2 \approx \frac{\sigma^2}{N}.$$

On the other hand, if the data rate is high enough that there are many measurements per integral correlation time, the measurements are not independent and the equivalent number of independent measurements is a function of how many correlation times are measured, *i.e.*

$$\sigma_N^2 \approx \frac{2\Lambda}{T}.$$

The average measurement rate, ρ, is approximately given by

$$\rho \approx \frac{N}{T}.$$

$$\sigma_N^2 \approx \frac{2\rho\Lambda}{N}.$$

We expect therefore that for low measurement rates the variance will approximate the expected variance for independent measurements. For high measurement rates, each measurement is not an independent estimate of the mean and we get something that approximates the result we obtained for continually sampled signals. With these approximations in mind, I will do the calculation more formally.

$$\langle (f_N - \langle f \rangle)^2 \rangle = \sigma_N^2 = \frac{1}{N^2} \langle \sum_k \sum_n \langle f_k f_n \rangle \rangle - \langle f \rangle^2.$$

Note that there are two sets of expected value signs in the expression. This is to signal the fact that we have to do the expected value over f and over the pairs of times, t_k and t_n. From 3.1.4, and 3.4.1, $RC(\tau)$ is the autocorrelation function of the underlying random process,

$$\sigma_N^2 = \frac{1}{N^2} \left\langle \sum_k \sum_n RC(t_k - t_n) \right\rangle - \langle f \rangle^2$$

$$\sigma_N^2 = \frac{\sigma^2}{N^2} \left\langle \sum_k \sum_n R(t_k - t_n) \right\rangle$$

$$\sigma_N^2 = \frac{\sigma^2}{N^2} \left(NR(0) + \left\langle \sum_{k \neq} \sum_n R(t_k - t_n) \right\rangle \right).$$

The probability of a particular time t_k is uniform, see 5.2.1.

Thus, we can compute the expected value of the variance of the mean,

$$\left\langle \sum_{k \neq} \sum_n R(t_k - t_n) \right\rangle = \frac{N(N-1)}{T^2} \int_0^T \int_0^T R(t_k - t_n) dt_k dt_n.$$

Repeating the calculation method used in 3.2.5, we get

$$\left\langle \sum_{k \neq} \sum_n R(t_k - t_n) \right\rangle = \frac{N(N-1)}{T} \int_{-T}^T (1 - \frac{|\tau|}{T}) R(\tau) d\tau \approx \frac{N(N-1)2\Lambda}{T}$$

$$\sigma_N^2 = \frac{\sigma^2}{N^2}\left(NR(0) + \frac{N(N-1)2\Lambda}{T}\right).$$

But, $(N-1)/T \approx \rho$, the average data rate and $R(0) = 1$. Thus,

$$\sigma_N^2 \approx \frac{\sigma^2}{N}(1 + 2\rho\Lambda). \tag{5.2.4}$$

The formula for the measurement variance is modified by random sampling in the manner we expected from the approximate calculations.

On the other hand, if I had posed the question in terms of measurement time instead of number of measurements, the problem would have looked somewhat different. In this case, we would be approximating the integral normally used in computing the time average of a continuous variable, *viz.*

$$\langle f \rangle = \frac{1}{T}\left\langle \int_0^T f(t)dt \right\rangle \frac{1}{T}\left\langle \sum_k f(t_k)(t_{k+1} - t_k) \right\rangle,$$

where T is the total time taken to obtain the measurements. This is equivalent to creating a new, filled, time series by simply filling in the empty spaces in a regularly spaced series with the last value obtained from the random sampling. It is usually called a "Sample and Hold" series.

Now we have to compute the expected value of the time between measurements as well as the expected value of f. In this case, we have assumed that the measurement statistics are independent of the function values, so we can do them separately. This simplifies the problem so we can immediately write

$$\langle f \rangle \approx \frac{1}{T}\langle f \rangle\left\langle \sum_k (t_{k+1} - t_k) \right\rangle = \frac{1}{T}\langle f \rangle T = \langle f \rangle.$$

The variance of the mean can be shown to be the same as 5.2.4, *viz.*

$$\sigma_T^2 \approx \frac{\sigma^2}{\rho T}(1 + 2\rho\Lambda) = \frac{\sigma^2}{N}(1 + 2\rho\Lambda).$$

5.2.2 *Spectrum and autocorrelation estimate using randomly sampled data*

Interestingly, some investigators involved in random sampling do not bother with the distinction between points in the derived series that contain data and those that do not. The new record is derived by dividing the data record in "slots" Δt long. The data is then scanned. If there is a measurement in a particular interval, it is recorded. If there is more than one measurement in the interval, just the first is recorded. If there is no measurement in the interval, a zero is recorded. In other words, the derived series $\{y_n\}$ is treated as if it were a data set taken at regular spacing, with data at each point. It is just that a lot of the points are zero. This is equivalent to considering the output to be the original data set $\{f_i\}$, imagined to be a complete regularly sampled time series, multiplied by a sampling series $S(i)$, where

$$S(i) = \sum_{k \varepsilon I} \delta_{ik} ,$$

where I is a random set of integers representing the times when a measurement was successfully taken.

$$\{y_i\} = \{f_i S(i)\} .$$

It is simple to show, using the same calculations as above, that if the sampling is independent of the value of the measurement, and you only use the non-zero channels,

$$\langle y \rangle = \langle f \rangle .$$

The formula for the measurement variance is also exactly the same as before, but the computation of the autocorrelation functions and the spectrum have some important differences, *viz.*

$$\langle R_{yy}(k) \rangle = \frac{1}{N} \left\langle \sum_i y_i y_{i+k} \right\rangle = \frac{1}{N} \left\langle \sum_i f_i f_{i+k} \right\rangle \langle S(i) S(i+k) \rangle ,$$

since the processes are independent.

$$\langle RC_{yy}(k) \rangle = \langle RC_{ff}(k) \rangle [\beta \delta_{0k} + \beta^2 (1 - \delta_{0k})] , \qquad (5.2.5)$$

where β is the probability of a measurement interval having a measured value in it. From above, (5.1.1)

$$\beta = (1 - e^{-\rho \Delta t}) .$$

If you ignore the first point, zero lag time, the expected correlation is still proportional to a faithful representation of the expected autocorrelation function of the original series for f.

The corresponding spectrum is given by

$$S_{yy}^2(m) = (\beta - \beta^2) R_{ff}(0) + \beta^2 S_{ff}^2(m) . \qquad (5.2.6)$$

There is a baseline introduced into the spectrum.

The behavior of the variance of the autocorrelation and the spectrum have been extensively studied by Gastor and Roberts for the randomly sampled system just described. The results they obtain for the measurement variances are:

$$\varepsilon^2\{R(p)\} \approx \frac{\sigma^4}{N_k}\left[\sum\nolimits_{j=1}^{N} R(j)R(p-j) + \frac{2}{\rho\sigma^2}R(p) + \frac{1}{\rho^2}\delta_0\right] \qquad (5.2.7)$$

and

$$\varepsilon^2\{S_{yy}^2(m)\} \approx \frac{N_s}{N_k}[S_{ff}^2(m) + \frac{\sigma^2}{\rho}]^2. \qquad (5.2.8)$$

N_k is the average number of non-zero pairs and N_s is the window width (see Chapter 3).

Random sampling, keeping all the data points in the sample results in an increased variance of the estimates of the autocorrelation functions and the spectra. Again, there is a natural maximum frequency induced by the measurement variance, not by the Nykvist criterion. It is beyond the scope of this text, but it can be shown that random sampling does make it possible to precisely compute the frequency of a pure sinusoid, when the frequency of the sinusoid is higher than the mean measurement rate.

You can take note of which slots are empty and compute the autocorrelation and spectrum in a slightly different way.

Typically, you take pains to see to it that the measurement interval for the second time series, Δt, is smaller than the micro-correlation time, but large enough to ensure most of the intervals have measurements in them. In what follows, we will assume that the interval is indeed shorter than the micro-correlation time of the series. Initially, the data will be treated as if each data point was taken exactly at the epoch t_i. Later, I will show the effect of relaxing this assumption.

The new series immediately generates three "N's"

— N = the number of channels in the series y.
— N_0 = the number of channels in y that contain at least one measurement.
— N_k = the number of pairs of channels k apart that are each non-zero.

From 5.2.2, the probability of *no* measurement in an interval is $e^{-\rho\Delta t}$.

The probability of at least one measurement in an interval is thus

$$1 - e^{-\rho\Delta t}.$$

The expected values for the last two N's are given by

$$\langle N_0 \rangle = N(1 - e^{-\rho \Delta t}) \qquad (5.2.9)$$

and

$$\langle N_k \rangle = N(1 - e^{-\rho \Delta t})^2 \,,$$

where Δt is the length of the sampling interval. This latter form comes about because the probability of a measurement in a given interval is completely independent of the probability for any other interval.

For simplicity, I will assume a zero mean signal.

The measured autocorrelation function is

$$\overline{RC}(k) = \frac{1}{N_k} \sum_{i=1}^{N_k} y_i y_{i+k} \,,$$

where, again, N_k is the number of non-empty pairs of entries k apart, in the derived series.

The expected autocorrelation function is

$$\langle \overline{RC}(k) \rangle = \frac{1}{N_k} \sum_{i=1}^{N} \langle y_i y_{i+k} \rangle = \frac{1}{N} \sum_{i=1}^{N-k} RC(k) = W(k, N) RC(k) \,.$$

The expected autocorrelation function is unbiased, if you divide the sums of the lag products by the actual number of non-zero pairs. The expected result is the same as for evenly spaced series with no zero entries. The only difference apparent at this point is the presence of the divisor N_k in the denominator, which is different for each lag time. However, it should be clear that except for N_0, the expected number for each lag time is the same, so that the computation can be simplified by dividing each sum of lag products by the average of the N_k.

The result does not appear, at this point, to put restrictions on Δt, so some have been known to claim that arbitrarily high frequencies can be measured using random sampling, whereas you are restrained to frequencies below the Nykvist frequency $(1/2\Delta t)$ for evenly spaced samples. We will examine this claim below.

Similarly, the measurement variance of the estimated autocorrelation function is

$$\varepsilon^2(k,l) = \frac{1}{N_k N_l} \sum_{j=1}^{N_k} \sum_{i=1}^{N_l} \langle y_i y_{i+k} y_j y_{j+l} \rangle - \langle \overline{RC}(k) \rangle \langle \overline{RC}(l) \rangle.$$

If we once again invoke Gaussian statistics to create a computable example, we get

$$\varepsilon^2(k,l) \approx \frac{\sigma^4}{N_k} \sum_{p=-N_k}^{N_k} W(k,N_k) W(l,N_l)$$
$$+ [R(p)R(p+k-l) + R(p-k)R(p+l)]. \qquad (5.2.11)$$

The result is exactly the same as before, except that the number of points used is N_k, which is less than N. As a consequence, any estimate of an autocorrelation function taken using data from a time period T with random sampling is always going to be noisier than one obtained using regular sampling, where every interval is non-empty.

Likewise, the measurement variance of the spectrum obtained from the autocorrelation function can be shown to be the same expression as before except N_k substitutes for N. When the necessary window is applied to the spectrum, the measurement variance is proportional to N_s/N_k. Recall that the number of points in the measured spectrum is proportional to N_s. Basically what this means is you want to have the same number of non-zero pairs, N_k, if you want to keep the same number of points in the spectrum and keep the same smoothness in the measured spectrum that is not randomly sampled.

Consider the situation where $\rho \Delta t \approx 1$. In this case, $N_k \approx 0.4N$. You would need roughly twice the number of points in y to get the same frequency resolution as for normal equally spaced sampling. If you decrease Δt by a factor of 4, then $N_k \approx 0.05N$. You would need about 20 times the number of points to get the frequency resolution belonging to Δt.

As you can see from the previous discussion, we may get any upper frequency we want, but we may have to measure for unreasonably long periods of time in order to be able to resolve spectral features at the high frequency end.

5.2.3 *Effect of finite sample intervals* Δt

Using a finite measurement interval Δt has the effect of averaging over that interval, since we record a measurement as occurring at the same epoch, no matter when it occurs in the interval. It can be shown that this has the effect of low-pass filtering the expected spectrum, *viz.*

$$\langle R(m)_{ran} \rangle = \int W(t', \Delta t) R(m\Delta t - t') dt' ,$$

thus

$$\langle S(p)_{ran} \rangle = W(p) S(p) .$$

Since $\hat{W}(\bullet)$ has a bandwidth of $\approx 1/\Delta t$, the sampling has the effect of placing an upper limit on the measurable spectrum. However, this should not cause any practical problem, since the highest useful frequency of most digital Fourier Transform routines is $1/2\Delta t$.

5.3 Sample and Hold Autocorrelation and Spectrum Estimation

This derivation closely follows the paper by Adrian and Yao (1987). I will assume here that the expected measurement rate is independent of the signal (velocity). For a sample and hold system, if you pick two epochs of time, t_1 and t_2 the velocity at those times are velocities that occurred at earlier times, say ξ_1 and ξ_2. If $t_2 \rangle t_1$, you actually have two possibilities. One, that no measurement occurs between the two times, in which case you get the same measurement at both times; the other is that you get at least one measurement between the two times. In all that follows, it is assumed that $u(t)$ is a stationary, zero mean time series. The latter assumption is made because we are interested in the fluctuations in the output. Non-zero means will be handled later. Using the tools developed above, we can write the probability of getting the velocity that occurred at ξ_1 at t_1 as follows:

$p(u(\xi_1), t_1) =$probability of a measurement in the interval $\xi_1, d\xi_1$ and no measurement in the interval $t_1 - \xi_1$.

$$p(u(\xi_1), t_1) = \rho e^{-\rho(t_1 - \xi_1)} d\xi_1 .$$

The probability of a measurement at ξ_2 appearing in the output of the sample and hold at t_2 from the interval between t_1 and t_2, is similarly given by

$$p((u(\xi_2), t_2) = \rho e^{-\rho(t_2 - \xi_2)} d\xi_2 .$$

The expression for the expected autocorrelation function of the sample and hold is given by

$$\langle R_{mm}(t_2, t_1) \rangle = \rho^2 \int_{-\infty}^{t_1} \int_{t_1}^{t_2} \langle u(\xi_1) u(\xi_2) \rangle e^{-\rho(t_1 - \xi_1)} e^{-\rho(t_2 - \xi_2)} d\xi_1 d\xi_2$$

$$+ \rho \int_{-\infty}^{t_1} \langle u^2(\xi_2) \rangle e^{-\rho(t_2 - \xi_2)} d\xi_2 . \tag{5.3.1}$$

We are able to write it this way because we have assumed that the sampling is independent of the value of the underlying series. The subscript "mm" denotes the sample and hold signal. Because $u(t)$ is stationary, we can write

$$R_{mm}(t_2 - t_1) = \rho^2 \int_{-\infty}^{t_1} \int_{t_1}^{t_2} R_{uu}(\xi_2 - \xi_1) e^{-\rho(t_1 - \xi_1)} e^{-\rho(t_2 - \xi_2)} d\xi_1 d\xi_2$$

$$+ \langle \sigma_u^2 \rangle e^{-\rho(t_2 - t_1)} .$$

Here, "uu" denotes the original signal. If R_{uu} is a known function, we can stop here. In general, it is not known and we need to do more computation to understand how this form of sampling affects the result.

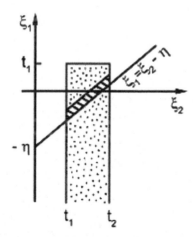

Fig. 5.1 Illustration of change of variable and ranges.

To do this, we need to change the form of the integration specified above. Noting that R_{uu} is a function of the difference $\xi_2 - \xi_1$, we do the initial integration on paths where this difference is constant. Define two variables $\eta = \xi_2 - \xi_1$ and $\tau = t_2 - t_1$. The initial integration is done in two regions, (see diagram). The first is for $\tau > \eta > 0$ and the second for $\infty > \eta \geq \tau$. The corresponding integrals are

$$\rho^2 \int_{t_1}^{t_1+\eta} e^{-\rho(t_1-\xi_2+\eta)} e^{-\rho(t_2-\xi_2)} d\xi_2$$

and

$$\rho^2 \int_{t_1}^{t_2} e^{-\rho(t_1-\xi_2+\eta)} e^{-\rho(t_2-\xi_2)} d\xi_2 \,.$$

The results of these integrations are respectively

$$\frac{\rho}{2}(e^{-\rho(\tau-\eta)} - e^{-\rho(\eta+\tau)}), \quad 0 < \eta < \tau$$

and

$$\frac{\rho}{2}(e^{-\rho(\eta-\tau)} - e^{-\rho(\eta+\tau)}), \quad \infty > \eta \geq \tau \,.$$

The two expressions can be combined to give an expression that covers the entire range of the variable η. The result of the inner integration is

$$\frac{\rho}{2}(e^{-\rho|(\tau-\eta)|} - e^{-\rho(\eta+\tau)}), \quad 0 < \eta < \infty \,. \tag{5.3.2}$$

We can now write the new expression for the expected measured autocorrelation function.

$$R_{mm}(\tau) = \langle \sigma_u^2 \rangle e^{-\rho\tau} + \frac{\rho}{2} \int_0^\infty R_{uu}(\eta)(e^{-\rho|\tau-\eta|} - e^{-\rho(\tau+\eta)}) d\eta \,. \tag{5.3.3}$$

This can be further modified to give a more familiar expression, *viz.*

$$R_{mm}(\tau) = \langle \sigma_u^2 \rangle e^{-\rho\tau} + \frac{\rho}{2} \int_{-\infty}^\infty R_{uu}(\eta) e^{-\rho|\tau-\eta|} d\eta \,. \tag{5.3.4}$$

From (3.3.7), the spectrum can be seen to be

$$S_{mm}(\omega) = \left(\frac{1}{1+\omega^2/\rho^2} \right) (S_{uu}(\omega) + 2\langle \sigma_u^2 \rangle/\rho) \,. \tag{5.3.5}$$

The spectrum of the sample and hold signal is the spectrum of the underlying process plus a "noise" caused by the switching levels, all multiplied by a low pass spectrum with a 3-db frequency of ρ.

If the signal is non-zero mean, simply add $\langle u \rangle^2$ to the expression above, 5.3.4.

5.4 Non-Independent Sampling

We are now in a position to examine situations where the probability of a measurement is a function of the measured value; $p(t_i|f_i)$ is not a constant.

The situation of interest is one where the measurement epochs are still random, but the conditional probability of a measurement depends on the value being measured. This situation is of particular interest in flow measurement using light scattering, where the measurements are obtained from the light scattered by small particles entrained in the flow. Often, the particle number density is so low that only one particle at a time is seen by the measurement system. Obviously, for a fixed particle number density, a speedier flow will carry more particles to the measurement region per unit time than will a less speedy flow. Thus you would expect a correlation between the measurement rate and the speed of the flow. The situation is complicated by other effects involved in the measurement process, such as the data logging system, but the chief difficulty in predicting what a given system will do is caused by the fact that you do not usually measure the speed of the flow at the measurement point, but one or more velocity components, with varying degrees of accuracy.

In any event, the probability of a measurement is a function of the quantity being measured. Optimally, you would like to have a processing scheme that made the results independent of the details of the dependence of the measurement rate on the quantity being measured. This section will run roughly parallel to the previous section. I will assume that f_i, the quantity being sampled is also a random variable. Ultimately, we want to know its statistics, the means, and second order statistics like the variance and the autocorrelation functions.

Here, the sampling is still random, but the probability of a measurement depends on the measured value. We must therefore express the probability of a measurement in terms of the measured value. Let the *conditional measurement rate* of a value f_i be denoted $\rho(f_i)$. This is the measurement rate you would get if the signal was maintained at f_i. It should be clear that the conditional probability of a measurement of f_i is given by (see 5.2.1)

$$p(t_i, f_i)dt_i = \frac{\rho(f_i)}{\langle \rho \rangle T}dt_i \,. \tag{5.4.1}$$

Again, we attempt to compute the expected value of the measurements.

$$\langle f_i \rangle = \left\langle \frac{1}{N} \sum_k f(t_k) \right\rangle$$

$$\langle f_i \rangle = \frac{1}{N} \sum_k \langle f(t_k) \rangle ,$$

but this time we have to be more careful calculating the expected value than we were before, since now the probability of a measurement is dependent on the value measured.

$$\langle f(t_k) \rangle = \frac{1}{T} \iint p(t_i, f_i) f_i dt_i df_i = \frac{1}{T} \iint p(t_i | f_i) p(f_i) dt_i df_i$$

$$\langle f(t_k) \rangle = \frac{1}{\langle \rho \rangle T} \iint \rho(f_i) p(f_i) f_i dt_i df_i .$$

$$\langle f(t_k) \rangle = \langle \rho(f) f \rangle / \langle \rho \rangle . \tag{5.4.2}$$

We can go no further without an explicit expression for $\rho(f)$. Assume for the moment that (unrealistically), $\rho(f) = \alpha f$, *i.e.*, the measurement rate is proportional to f.

Then,

$$f(t_k) = \langle \rho(f) f \rangle / \langle \rho \rangle = \langle f^2 \rangle / \langle f \rangle = (\langle f \rangle^2 + \sigma^2) / \langle f \rangle .$$

There is a bias error of $\sigma^2 / \langle f \rangle$. As I mentioned above, this case is a rather artificial one, in that, for real situations, the dependence of the rate on f is more complex. However, it gives some idea of the kind of problem you can get if you encounter a situation where the probability of a measurement is a function of the measurement itself. I emphasize that if the exact form of the dependence of the measurement probability as a function of the measurement is known, *the bias can be eliminated to any arbitrary degree of accuracy.* Let $R(f_i)$ be the known conditional probability of a measurement of f_i, then the unbiased average can be shown to be given by

$$\langle f_i \rangle = \frac{\sum f_i / R(f_i)}{\sum 1 / R(f_i)} .$$

The problem is that the form of the measurement probability is usually not known precisely or the information needed to compute it is incomplete. In the latter case, you can make matters worse by attempting to correct the data inappropriately.

If your data set is large enough, you can attempt to measure $R(f_i)$. Set an appropriately small range for the difference in $f_i, \Delta f_i$. Pick a value for f_i. Search for all values of f that are within Δf_i of the chosen value. Then search for the number of measurements in a time Δt after each occurrence. Denote this number $n(f_i)$. The conditional probability is then estimated from

$$R(f_i) \propto n(f_i)/n_{\text{total}}.$$

If possible, a better strategy is to find a method of treating the data that is insensitive to the exact form of the conditional measurement probability. In this way, at least the effect of the conditional probability can be minimized.

5.4.1 *Sample and hold (rate depends on f)*

Consider now the situation where we attempt to approximate the time average by using a Sample and Hold process, when the relative measurement rate depends on the process.

$$\langle f \rangle = \frac{1}{T} \left\langle \int_0^T f(t)dt \right\rangle$$

$$\approx \frac{1}{T} \left\langle \sum_k f(t_k)(t_{k+1} - t_k) \right\rangle = \frac{1}{T} \sum_k \langle f(t_k)(t_{k+1} - t_k) \rangle$$

$$\langle f \rangle \approx \frac{1}{T} \sum \int f p(f) \frac{\rho_f}{\langle \rho \rangle} \langle \Delta T | f \rangle df.$$

Since $\langle \rho \rangle = N/T$,

$$\langle f \rangle \approx \frac{1}{N} \sum \int f p(f) \rho_f \langle \Delta T | f \rangle df = \langle f \rho_f \langle \Delta T | f \rangle \rangle.$$

Now, it should be more obvious that if $\rho_f \langle \Delta T | f \rangle = 1$, that the measurements would be unbiased, since the expression for the expected value would reduce to $\langle f \rangle$.

Since the probability of a value being measured is coupled to the possible length of the interval following that measurement, we have to find an expression for the time between measurements as a function of the value of the function at the beginning of the interval, *i.e.*, the conditional inter-arrival time distribution. The inter-arrival time distribution is computed from the probability of no measurements in the time interval ΔT and the probability of a measurement in the differential time interval at the end of ΔT, $d\Delta T$.

$p(\Delta T) = $ (Probability of no measurement in ΔT) times the probability of a measurement in the interval $d\Delta T$.

$p(\Delta T) = $ Exp $[-$expected number of measurements in $\Delta T]\rho d\Delta T$.

In other words, we need to compute the expected number of measurements following f_i. This problem is complicated by the fact that we are sampling a random variable, albeit one with some correlation. It is not possible to derive an exact, general result, since the solution depends on the details of the experiment. The following heuristic treatment should give some idea of the kind of behaviors this kind of sampling can have.

The expected number of measurements in an interval ΔT long, if the value at the beginning of the interval is f, is given by

$$\langle n(\Delta T|f)\rangle = \left\langle \int_0^{\Delta T} \rho(f(t))dt \right\rangle = \int_0^{\Delta T} \langle \rho(f(t))\rangle dt \,,$$

where it is understood that $f(0)$ is f.

If ΔT is short compared to Λ, the underlying processes correlation time, $\langle n(\Delta T)\rangle \approx \rho_i \Delta T$. For intervals that are long compared to Λ, $\langle n(\Delta T)\rangle \approx \langle \rho \rangle \Delta T$. The transition between the two behaviors should occur sometime around the correlation time. Defining $\Delta \rho_i = \rho_i - \langle \rho \rangle$, a plausible form for the conditional expected value of $\Delta \rho_j$ can be written

$$\langle \Delta \rho_j | \Delta \rho_i \rangle = \Delta \rho_i R_\rho(j - i) \,,$$

where $R_\rho(i - j)$ is the covariance of $\rho(f)$ at lag time $j - i$. This expression is true for any function of f_i. For this case, the conditional inter-arrival time distribution can be written

$$p(\Delta T|f)d\Delta T = \langle \rho_j | \rho_f \rangle \text{Exp}\left[-\int_0^{\Delta T} \langle \rho_j | \rho_f \rangle dt \right] d\Delta T$$

$$= (\langle \rho \rangle + \Delta \rho_f R_\rho(\Delta T))\text{Exp}\left[-\langle \rho \rangle \Delta T + \Delta \rho_i \int_0^{\Delta T} R_\rho(t)dt \right]$$

$$+ \, d\Delta T \,.$$

In order to make this expression more amenable to numerical manipulation, I rewrite it as

$$p(\tau|f_i)d\tau = \langle\rho\rangle\Lambda\left(1 + (\frac{\rho_i}{\langle\rho\rangle} - 1)R_\rho(\tau)\right)$$

$$\times \text{Exp}\left[-(\langle\rho\rangle\Lambda(\tau + \left(\frac{\rho_i}{\langle\rho\rangle} - 1\right)\int_0^\tau R_\rho(t)dt))\right]d\tau,$$

where $\tau = \Delta T/\Lambda$. The parameter $\langle\rho\rangle\Lambda$ is the expected number of measurements per correlation time. We can get some idea of the behavior of the system by making the reasonable assumption that

$$R_\rho(\tau) \simeq e^{-\tau}.$$

If the underlying process is turbulent, the time scale of the fluctuations in the rate should be essentially the same as the time scales of the underlying turbulence.

Using this approximation, we can now compute estimates of $\langle\Delta T|f\rangle$ numerically. Table 5.1. Note that if the expected time between measurements given the initial measurement were the inverse of the initial measurement rate, ρ_f, the product $\rho_f\langle\tau|f\rangle$ would be one.

Table 5.1

$\rho_i/\langle\rho\rangle$	$\langle\rho\rangle\Lambda = 0.1$	$\langle\rho\rangle\Lambda = 1$	$\langle\rho\rangle\Lambda = 10$
0.6	0.62	0.74	0.91
0.8	0.81	0.88	0.97
1.0	1.0	1.0	1.0
1.5	1.43	1.18	1.02
2.0	1.83	1.26	1.03

The column to the left is $\rho_i\langle\rho\rangle$. The rest of the columns are the values of the product of the conditional inter-arrival time multiplied by the conditional measurement rate.

From Table 5.1, it can be seen that a Sample and Hold system does indeed take on the behavior that the product of the conditional measurement rate and the conditional inter-arrival rate becomes one at higher values of the measurements per correlation time. Closer examination of the simulation results show that the transition to unbiased behavior starts at approximately $\langle\rho\rangle\Lambda = 5$. This behavior is insensitive to the form of $\rho(f_i)$.

It merely requires that the probability be high for getting a measurement in a time that is short compared to the correlation time of the underlying process, Λ. Also, many situations are encountered where the range of variation of the measurement rate is relatively small, on the order of 5–10%. In the latter situation, the distortion of averages is negligible for sampling rates higher than 5 measurements per correlation time.

5.4.2   *Controlled sampling*

Another strategy for attempting to get unbiased estimates of the statistics of a system where the random samples are conditioned by the values, is to attempt to make the sampling of the data set less conditioned on the measurements. Consider making another time series that is sampled at even intervals, δt

As before, we place the first value from the original series into the interval and zero if there is no value corresponding to that time interval in the original series. A variation is to average the values encountered in that interval if multiple values are encountered. Obviously, we wish to pick δt smaller than the process micro time scale. The probability of getting at least one measurement of measurement f_i in the interval is given by

$$p_m(t_i|f_i) = (1 - e^{-\rho_i \delta t})/\langle 1 - e^{-\rho_i \delta t}) \rangle .$$

The expected value of the new series is given by

$$\langle f \rangle \approx \frac{1}{\langle (1 - e^{-\rho_i \delta t}) \rangle} \int (1 - e^{-\rho(f_i)\delta t}) f_i p(f_i) df_i .$$

If $\rho \delta t \gg 4$, the measurement probability in the new series is no longer conditioned on the value of the measurement. Under these conditions, the sampling method looks like a good one. The probability of having an empty channel is very low. In many situations, the measurement rate is too low to meet this criterion, especially given that the micro time can be at least an order of magnitude smaller than the macro time scale, Λ. If, however, the measurement rate is high enough, using this method of sampling effectively creates a system where the probability of the appearance of a value in the output is not conditioned by the value.

On the other hand, if you try using the derived series to estimate second order statistics like the autocorrelation function, you can get bizarre results.

Let the derived series be denoted y_i and let the statistics for f_i be Gaussian. The autocorrelation of the derived series can be computed as shown below.

5.4.3 *Autocorrelation estimation*

$$\langle y_i^2 \rangle = \frac{\langle (1 - e^{-\rho_i \delta t}) f_i^2 \rangle}{\langle (1 - e^{-\rho_i \delta t}) \rangle} = \tilde{R}(0) \,.$$

$$\langle y_i y_j \rangle = \frac{\langle (1 - e^{-\rho_i \delta t})(1 - e^{-\rho_j \delta t}) f_i f_j \rangle}{\langle (1 - e^{-\rho_i \delta t}) \rangle \langle (1 - e^{-\rho_i \delta t}) \rangle} = \tilde{R}(j - i) \,.$$

Let $j - i$ be small, say 1, and let ρ_i be such that the measurement rate is proportional to f_i. Further, let ρ_i be small enough that the conditional probability of a measurement can be approximated by $\rho_i / \langle \rho \rangle$.

Then

$$\tilde{R}(0) \simeq \frac{\langle f^3 \rangle}{\langle f \rangle} = \langle f \rangle^2 + 3\sigma^2 \,,$$

$$\tilde{R}(1) \approx \frac{\langle f^4 \rangle}{\langle f \rangle^2} = \langle f \rangle^2 + 6\sigma^2 + \frac{3\sigma^4}{\langle f \rangle^2} \,.$$

The first point of an unbiased autocorrelation function should be $\langle f \rangle^2 + \sigma^2$. Thus the initial point is biased. But, more important, note that the second point of the autocorrelation function is larger than the first! This cannot be for a proper autocorrelation function, otherwise you have to believe that some point correlates better with another data point than with itself. What has happened is that we did not compute a proper autocorrelation function. If you examine the process carefully, you will see that we had a distinct bias toward pairs of data points with a larger value than toward pairs with a smaller value. We thus got more large products in the summation than small ones.

On the other hand, if $\rho_i \delta t$ is large, the probability of a measurement is not related to the underlying signal and the measured autocorrelation function is unbiased.

5.5   Photon Detection

In doing an experiment involving light detection, you will always experience fluctuations due to the fact that you are detecting photons, which are discrete events. This will be true whether you are directly photon counting or are using an analog detector. In what follows, we will assume that the light source is a coherent one such as a laser and that the source of the light being detected is fluctuating.

The statistics of light from a coherent source are Poisson. One can predict the average number of photons that will arrive in a given time, but you cannot predict exactly how many will arrive or when the photons will be detected. The statistics of light from a coherent source are Poisson. The probability of measuring k photons in a time interval Δt, if the expected number of photons in the interval is μ, is given by

$$p(k) = e^{-\mu} \frac{\mu^k}{k!} \, .$$

In an analog system, Δt can be taken to be the inverse of the bandwidth of the system.

5.5.1   *Moments*

If μ is a function of time, you are dealing with what is known as a *doubly stochastic Poisson point process*. It is instructive to compute the moments of the measured signal.

Let n be the measured signal.

$$\langle n \rangle = \left\langle \sum_{k=0}^{\infty} k e^{-\mu} \frac{\mu^k}{k!} \right\rangle = \langle \mu \rangle$$

$$\langle n^2 \rangle = \left\langle \sum_{k=0}^{\infty} k^2 e^{-\mu} \frac{\mu^k}{k!} \right\rangle = \langle \mu^2 \rangle + \langle \mu \rangle \, .$$

Both of these were proven in Chapter 1. The second also shows that the variance of the measurements has an additional component, *i.e.*

$$\sigma_n^2 = \sigma_\mu^2 + \langle \mu \rangle \, .$$

Some higher moments are of interest to us for computing the expected autocorrelation functions and the variance of estimated autocorrelation functions.

$$\langle n^3 \rangle = \langle \mu^3 \rangle + 3\langle \mu^2 \rangle + \langle \mu \rangle$$

$$\langle n^4 \rangle = \langle \mu^4 \rangle + 6\langle \mu^3 \rangle + 7\langle \mu^2 \rangle + \langle \mu \rangle$$

$$\langle n_1 n_2 \rangle = \langle \mu_1 \mu_2 \rangle$$

$$\langle n_1 n_2 \ldots n_q \rangle = \langle \mu_1 \mu_2 \ldots \mu_q \rangle$$

$$\langle n_1^2 n_2 \rangle = \langle \mu_1^2 \mu_2 \rangle + \langle \mu_1 \mu_2 \rangle$$

$$\langle n_1^2 n_2^2 \rangle = \langle \mu_1^2 \mu_2^2 \rangle + \langle \mu_1^2 \mu_2 \rangle + \langle \mu_1 \mu_2^2 \rangle + \langle \mu_1 \mu_2 \rangle$$

$$\langle n_1^3 n_2 \rangle = \langle \mu_1^3 \mu_2 \rangle + 3\langle \mu_1^2 \mu_2 \rangle + \langle \mu_1 \mu_2 \rangle$$

$$\langle n_1^2 n_2 n_3 \rangle = \langle \mu_1^2 \mu_2 \mu_3 \rangle + \langle \mu_1 \mu_2 \mu_3 \rangle$$

5.5.2 *Autocorrelation estimation*

These formulae are useful for calculating the expected measured autocorrelation function and the variance of the measured autocorrelation. Assuming we take the data in regular intervals and if μ is a stochastic variable,

$$R(k) = \frac{1}{N} \sum_{j=1}^{N-k} n(j)n(j+k)$$

$$\langle R(k) \rangle = \frac{1}{N} \sum_{j=1}^{N-k} \langle n(j)n(j+k) \rangle$$

$$= \frac{1}{N} \sum_{j=0}^{N-k} (\langle \mu(j)\mu(j+k) \rangle + \delta_{0k}\langle \mu \rangle)$$

$$= W(k,N)R_{\mu\mu}(k) + \delta_{0k}\langle \mu \rangle \,.$$

This is essentially the same result we got in Chapter 3 for regularly sampled non-photon signals, except for the added term that appears at $k = 0$. This is often referred to as the "shot" term.

The covariance is a bit more complex, *viz.*

$$\Lambda_{mn} = \left\{ \begin{array}{l} \dfrac{1}{N^2} \sum_{q=-(N-n-1)}^{N-m-1} C(q,m,n) \times [\langle \mu(0)\mu(q)\mu(q+m)\mu(n) \rangle \\[4pt] + \delta_{0n} \langle \mu(0)\mu(q)\mu(q+m) \rangle + \delta_{0m} \langle \mu(0)\mu(q)\mu(n) \rangle \\[4pt] + \delta_{0m}\delta_{0n} \langle \mu(0)\mu(q) \rangle] \end{array} \right\}$$

$$+ \frac{N-m}{N^2} \times [\langle \mu(0)\mu(m)\mu(n) \rangle + \langle \mu(0)\mu(n)\mu(n-m) \rangle]$$

$$+ \frac{N-(m+n)}{N^2} \times [\langle \mu(0)\mu(-m)\mu(n) \rangle + \langle \mu(0)\mu(n)\mu(n+m) \rangle]$$

$$+ \frac{N-n}{N^2} \times [2\delta_{0m} \langle \mu(0)\mu(n) \rangle + \delta_{mn} \langle \mu(0)\mu(n) \rangle]$$

$$+ \frac{2N-(m+n)}{N^2} \delta_{0n} \langle \mu(0)\mu(m) \rangle$$

$$+ \frac{1}{N} \delta_{0m}\delta_{0n}[\langle \mu^2 \rangle + \langle \mu \rangle] - \langle \mu(0)\mu(m) \rangle \langle \mu(0)\mu(n) \rangle$$

Here,

$$C(q,m,n) = \begin{cases} N+q-n, & (N-n-1) \le q \le -1 \\ N-n, & 0 \le q \le n-m \\ N-q-m, & (n-m-1) \le q \le (N-m-1) \end{cases}.$$

For large N, $C(q,m,n) = N$.

References

1. "Photoelectron Statistics: With Applications to Spectroscopy and Optical Communications," B. Saleh, Springer-Verlag, New York 1968.
2. "The spectral analysis of randomly sampled records by a direct transform," Gaster and Roberts, *Proc. R. Soc. London.* **A354**, 27–58 (1977).
3. "Power spectrum of fluid velocities measured by laser Doppler velocimetry," R. J. Adrian and C. S. Yao, *Experiments in Fluids* **5**, 17–28 (1987).
4. "A maximum likelihood based autocorrelation processor for a photon resolved laser Doppler anemometer," P. M. Howard and Robert V. Edwards, *Meas. Sci. Techn.* **7**, 801–822 (1996).

Index